RNAS

W9-CRL-638

ACS SYMPOSIUM SERIES **830**

Biological Systems Engineering

Mark R. Marten, Editor
University of Maryland Baltimore County

Tai Hyun Park, Editor
Seoul National University

Teruyuki Nagamune, Editor
University of Tokyo

American Chemical Society, Washington, DC

Chemistry Library

Library of Congress Cataloging-in-Publication Data

Biological systems engineering / Mark R. Marten, Tai Hyun Park, Teruyuki Nagamune, editors.

    p. cm.—(ACS symposium series ; 830)

    Includes bibliographical references and index.

    ISBN 0–8412–3796–4

    1. Biochemical Engineering—Congresses. 2. Systems Engineering—Congresses.

    I. Marten, Mark R., 1964-. II. Park, Tai Hyun, 1957- III. Nagamune, Teruyuki, 1952- IV. Series.

TP248.3 .B56   2002
660.6′3—dc21
                                  2002025576

The paper used in this publication meets the minimum requirements of American National Standard for Information Sciences—Permanence of Paper for Printed Library Materials, ANSI Z39.48–1984.

Copyright © 2002 American Chemical Society

Distributed by Oxford University Press

All Rights Reserved. Reprographic copying beyond that permitted by Sections 107 or 108 of the U.S. Copyright Act is allowed for internal use only, provided that a per-chapter fee of $22.50 plus $0.75 per page is paid to the Copyright Clearance Center, Inc., 222 Rosewood Drive, Danvers, MA 01923, USA. Republication or reproduction for sale of pages in this book is permitted only under license from ACS. Direct these and other permission requests to ACS Copyright Office, Publications Division, 1155 16th St., N.W., Washington, DC 20036.

The citation of trade names and/or names of manufacturers in this publication is not to be construed as an endorsement or as approval by ACS of the commercial products or services referenced herein; nor should the mere reference herein to any drawing, specification, chemical process, or other data be regarded as a license or as a conveyance of any right or permission to the holder, reader, or any other person or corporation, to manufacture, reproduce, use, or sell any patented invention or copyrighted work that may in any way be related thereto. Registered names, trademarks, etc., used in this publication, even without specific indication thereof, are not to be considered unprotected by law.

PRINTED IN THE UNITED STATES OF AMERICA

TP
248
.3
B56
2002
CHEM

Foreword

The ACS Symposium Series was first published in 1974 to provide a mechanism for publishing symposia quickly in book form. The purpose of the series is to publish timely, comprehensive books developed from ACS sponsored symposia based on current scientific research. Occasion-ally, books are developed from symposia sponsored by other organizations when the topic is of keen interest to the chemistry audience.

Before agreeing to publish a book, the proposed table of contents is reviewed for appropriate and comprehensive coverage and for interest to the audience. Some papers may be excluded to better focus the book; others may be added to provide comprehensiveness. When appropriate, overview or introductory chapters are added. Drafts of chapters are peer-reviewed prior to final acceptance or rejection, and manuscripts are prepared in camera-ready format.

As a rule, only original research papers and original review papers are included in the volumes. Verbatim reproductions of previously published papers are not accepted.

ACS Books Department

Contents

Microbial and Enzymatic Systems

Animal Cell Systems

Combinatorial Bioengineering
and Analytical Methods

Indexes

Preface

What many think of as modern biochemical production technology began in the early 1940s, as an effort to meet the increasing demand for penicillin during World War II. During that time, penicillin production moved from surface culture in milk bottles to submerged culture in tanks as large as 100,000 L. In the 60 years since then, significant advances in biochemical production technologies have occurred, as dramatic advances have been made in our understanding of cellular systems, and how these systems can be manipulated to produce desirable products. Today, production of biochemicals includes not only traditional subjects such as fermentation engineering and applied microbiology, but subjects as diverse as recombinant DNA technology, metabolic engineering, genomics, and combinatorial chemistry (just to name a few). This field continues to evolve as new technologies become available to operate, monitor and control bioprocesses, and as new biological discoveries are made at an ever increasing pace.

The majority of the papers contained in this text were presented at the PacifiChem 2000 conference, held in Honolulu, Hawaii in December 2000. Our goal in organizing this symposium and in compiling this book was to bring together scientists from around the Pacific Rim who are using various modern biological tools to study various aspects of biochemical production. The book is divided into three sections: (1) microbial and enzymatic systems, (2) animal cell systems, and (3) combinatorial engineering and analytical methods. Expression systems discussed in various chapters are used for the production of organic acids, specialty sugars, enzymes, and antibodies. Our hope is that this book will benefit the reader by providing a glimpse of how modern biological tools are being applied to the production of a mixture of biochemical products.

The editors thank the American Chemical Society Books Department of the Publications Division for assistance in the process, especially Kelly Dennis and Stacy VanDerWall in acquisitions and Margaret Brown in editing/production for all their help.

Mark R. Marten
University of Maryland – Baltimore County
Department of Chemical and Biochemical Engineering
1000 Hilltop Circle
Baltimore, MD 21250

Tai Hyun Park
Seoul National University
School of Chemical Engineering
San 56–1, Shilim-dong, Kwanak-gu
Seoul 151–744, Korea

Teruyuki Nagamune
The University of Tokyo
Department of Chemistry and Biotechnology
School of Engineering
7–3–1 Hongo, Bunkyo-ku
Tokyo, Japan 113–8656

Chapter 1

Biological Systems Engineering: An Overview

Mark R. Marten[1], Tai Hyun Park[2], and Teruyuki Nagamune[3]

[1]Department of Chemical and Biochemical Engineering, University
of Maryland Baltimore County, Baltimore, MD 21250
[2]School of Chemical Engineering, Seoul National University,
Seoul 151–744, Korea
[3]Department of Chemistry and Biotechnology, Graduate School
of Engineering, The University of Tokyo, Tokyo 113–8656, Japan

Biological systems have been used for thousands of years to produce a wide variety of products. These historical processes primarily involved microbial fermentation, and were used to produce food and alcohol. In the last 50 years, scientific advances have allowed the use of a wide rage of biological systems, that include not only microbial fermentation, but culture of animal and insect cells, use of enzymatic systems, and combinatorial biochemistry. These approaches have been employed to produce a broad range of products that span the spectrum from high value pharmaceuticals to commodity chemicals. This chapter gives an overview of some of biological systems currently used, the engineering of these systems, and the products they are used to produce.

Microbial and Enzymatic Systems

With the introduction of recombinant DNA technology, a completely new approach to the development of biochemical processes became possible. Various genetically modified microorganisms are currently used for the production of not only recombinant proteins but also non-protein compounds. The recombinant protein products include vaccines, therapeutic agents, and animal growth hormone. Recombinant DNA techniques have also enhanced the production of a broad range of low-molecular-weight non-protein compounds including amino

© 2002 American Chemical Society

acids, antibiotics, vitamins, and dyes. Genetically modified microorganisms are also used to produce biopolymers that are useful in the food-processing, manufacturing, and pharmaceutical industries. Various kinds of microorganisms are currently being used as biological factories for the production of such commercial products. Among the microorganisms, *Escherichia coli* and *Saccharomyces cerevisiae* have been widely used as host microorganisms, since *E. coli* is the most studied organism and *S. cerevisiae* is a well-studied nonpathogenic eukaryotic microorganism.

For the production of commercial products, microbes should be cultivated in large quantities under conditions that give maximal productivity. This requires the development of high-expression host cell/vector systems, which can be regulated and are stable in large-scale processes. High cell density culture is one way to increase productivity. In high cell density culture, one major problem is the formation of undesirable by-products. For example, in *E. coli* culture acetic acid is a major inhibitory by-product to cell growth, and is produced when cells are grown under oxygen-limiting conditions and in the presence of excess glucose. The best way to obtain high cell density is to use a fed-batch fermentation strategy, and as a result fed-batch operation is one of the most popular modes of operation for the industrial production of recombinant proteins. Various feeding strategies for fed-batch operations have been proposed to obtain a high cell density and high concentration of recombinant proteins. These include feeding strategies using pH-stat, DO-stat, glucose-stat, and controlling the specific growth rate.

Gene manipulations either create a new pathway or augment a preexisting pathway for the production of a specific compound. The production ability of microorganisms could be improved more systematically by metabolic engineering (*1*). In contrast to the prior focus on single reactions in a pathway, metabolic engineering concerns the enumeration and quantification of intracellular metabolic fluxes, and emphasizes the integrated metabolic networks. This systematic and integrated approach can increase product yield and productivity, and can also provide the formation of novel biochemical products. More global cellular regulatory data are accumulating with the current development of functional genomic and proteomic technologies. This will provide more detailed information for the biochemical production.

Enzymatic systems have also been widely used in a large variety of industrial applications such as production of high-fructose syrups from corn starch, enzymatic hydrolysis of lignocelluloses to monomeric sugars, manufacturing of L-amino acids by resolution of racemic amino acid mixtures, manufacture of semi-synthetic penicillins and even production of commodity chemicals like acryl amide. Important aspects of enzymatic system development are to obtain suitable enzyme with high activity and stability, to improve the properties of enzyme, and to design optimal bioreactor system for enzymatic reaction and product separation. As for screening of enzymes, microorganisms in extreme environments such as high temperature, salt concentration, pressure and alkaline pH, have recently drawn much attention as good sources, as enzymes from these microorganisms possess higher stability and broader substrate specificity compared to those from microorganisms from moderate environments.

Artificial evolution techniques, or so called "directed evolution" has been utilized to improve the properties of a number of enzymes *(2)*. This method has emerged as a powerful alternative to rational approaches for enzyme engineering. Directed enzyme evolution typically begins with the creation of a library of mutated enzyme genes, which are created using mutagenesis methods such as error-prone PCR, oligonucleotide-cassette mutagenesis and DNA-shuffling. Selection or rapid screening is then used to identify improved enzymes with respect to the desired property. The genes encoding these enzymes are subsequently subjected to further cycles of random mutagenesis, recombination and screening in order to accumulate beneficial mutations. As a result of enzyme screening from extremophiles, or directed enzyme evolution, enzymes with high activity, thermostability, substrate specificity, enantioselectivity and organic solvent tolerance have been obtained. These enzymes will contribute to the development of green chemical process.

Animal Cell Systems

Although microbial systems offer many advantages for the production of biochemical products, they have limitations in the production of glycoproteins. Many animal cell systems are able to glycosylate proteins in a fashion that is reasonably similar to the way in glycosylation occurs in a human being. Biological properties of the glycoproteins may depend on the glycosylation pattern, which also affects the therapeutic activity of pharmaceutical glycoproteins. Therefore, many pharmaceutical glycoproteins are produced in recombinant animal cell culture. Mammalian cell and insect cell systems have been widely used for the production of glycoproteins.

The insect cell-baculovirus system has received rapid and wide acceptance as an alternative to classical bacterial or yeast systems for the production of recombinant proteins. This system has several advantages, including high expression owing to a strong polyhedrin promoter, production of functionally and immunogenetically active recombinant proteins due to proper post-translational modifications, and nonpathogenicity of the baculoviruses to vertebrates and plants (3).

The key issues of the animal cell system have been concerned with the development of high-expression host cell/vector systems, the high cell density culture technology on a large-scale, the development of an efficient process for the production of recombinant proteins or viruses, and media development including low serum and serum-free media.

For the efficient production of cloned-gene protein, the high rate of cloned-gene expression and the viability of host cell are important factors. The increased host cell longevity favors the longer production of recombinant proteins, which results in higher productivity. Cell death can occur by either necrosis or apoptosis. This depends on the level of stress experienced by the cells. In animal cell culture, the decrease in cell viability is a consequence of apoptosis, which is a feature of many commercially important animal cell lines. Apoptosis is triggered by infection with viral vectors, suboptimal cultivation conditions, or heterologous gene expression, resulting in the lower process productivity (4). To suppress apoptosis in cell culture processes, three basic approaches have been conducted. Those are elimination of nutrient deprivation, addition of apoptosis-inhibiting compounds and over expression of apoptosis suppressor genes.

Combinatorial Bioengineering and Analytical Methods

Since the early 1990's, one of the key goals of combinatorial chemistry has been to revolutionize drug discovery. In the time that has elapsed, combinatorial methods for generating molecular libraries, coupled with high-throughput screening, have become core technologies for the identification of ligands to receptors and enzymes. The identified ligands can be utilized as powerful tools for pharmacological studies, and are essential as lead compounds for drug development.

Combinatorial libraries of small organic molecules can be generated by a variety of synthetic methods. For example, combinatorial libraries composed of completely random sequences of peptides or oligonucleotides can be synthesized by solid-phase synthesis with mixtures of activated amino acids or nucleotides in randomized coupling. Libraries consisting of random, site-directed mutants of a specific protein or nucleic acid oligonucleotide are composed of many variants of an initial parental molecule and are generated by biological machinery such as cells, phage, PCR and cell free protein synthesis systems. The phage display, cell surface display and ribosome display systems can provide combinatorial libraries with a pair of DNA and corresponding peptide or protein.

In any combinatorial library composed of peptides or oligonucleotides, all structures are built from a common set of chemical building blocks, with each molecule possessing a unique combination or sequence of these building blocks at each synthetically incorporated position. Additionally, the molecules all possess a common structural core or synthetic linkage, dictated by the type of molecules in the library and by the actual synthetic strategy employed. For example, collections of peptides or protein molecules in a combinatorial library are usually built from the 20 naturally occurring amino acids, and possess a common synthetic linkage, an amide bond between each position in the polymeric molecule. Thus the synthesis of combinatorial libraries composed of peptides or oligonucleotides can be done in high-throughput and systematic fashion.

Combinatorial approaches have been most successful when a semi-rational approach has been used to design the library of molecules to be prepared and tested. In these efforts, libraries are designed by using knowledge of the mechanism or structure of the biological target, or by basing the library upon lead compounds that have previously been identified to bind to the biological target. Unfortunately, structural or mechanistic information for many biological targets is either unavailable, or does not provide sufficient insight to enable productive library design. Additionally, for many targets, lead compounds have not yet been identified or novel motifs for binding are desired. Thus, the route from design and synthesis of compound libraries to identification of biologically valid lead structure is still long and tedious. To overcome this situation, highly parallel approaches, including system automation, are required not only in synthesis but also in analysis and screening. Creating an efficient interface between combinatorial syntheses and bioassay is indispensable to satisfy these demands. For example, screening in the interior of resin beads, that of microarrays and label-free affinity binding detection have been developed *(5)*.

References

1. Stephanopoulos, G.; Aristidou, A. A.; Nielsen, J. *Metabolic Engineering: Principles and Methodologies;* Academic Press: San Diego, CA, 1998.
2. Kuchener, O.; Arnold, F. H. *Directed evolution of Enzyme Catalysts*; *TIBTECH*, **1997**, *15*, 523-530.

6

3. Shuler, M. L.; Wood, H. A.; Granados, R. R.; Hammer, D. A. *Baculovirus Expression Systems and Biopesticides;* Wiley-Liss: New York, NY, 1995.
4. Kumar, S. *Apoptosis: Biology and Mechanisms;* Springer: New York, NY, 1999.
5. Rademann, J.; Jung, G. *Integrating Combinatorial Synthesis and Bioassay; Sience,* **2000**, *287*, 1947-1948.

Microbial and Enzymatic Systems

Chapter 2

Metabolic Systems Engineering Approach for Efficient Microbial Fermentation and Future Perspectives

Kazuyuki Shimizu

Department of Biochemical Engineering and Science, Kyushu Institute of Technology, Iizuka, Fukuoka 820–8502, Japan

The recent advancement on metabolic systems engineering was reviewed based on our recent research results in terms of (1)metabolic signal flow diagram approach, (2)metabolic flux analysis (MFA) in particular with isotopomer distribution using NMR and/or GC-MS, (3)synthesis and optimization of metabolic flux distribution (MFD), (4)modification of MFD by gene manipulation and by controlling culture environment, (5)metabolic control analysis (MCA), (6)design of metabolic regulation structure, and , (7) identification of unknown pathways with isotope tracing by NMR. The main characteristics of metabolic engineering is to treat metabolism as a network or entirety instead of individual reactions. The applications were made for poly-3-hydroxybutyrate (PHB) production using *Ralstonia eutropha* and recombinant *Escherichia coli,* lactate production by recombinant *Saccharomyces cerevisiae,* pyruvate production by vitamin auxotrophic yeast *Toluropsis glabrata,* lysine production using *Corynebacterium glutamicum,* and energetic analysis of photosynthesic microorganisms. The characteristics of each approach were reviewed with their applications. It is becoming important to study gene and protein expressions in relation to metabolic regulations.

Introduction

The recent progress in biotechnology enables us to improve the quality of our lives and environment, and its research field has been expanded from basic science to engineering such as (1) bioinformatics including functional genomics and proteomics, (2) protein engineering including protein structure - function and structure – activity relationships etc., (3) recombinant techniques including random mutation, DNA shuffling, and phage – display technique, (4) metabolic engineering, and (5) bioprocess engineering [1].

   © 2002 American Chemical Society

In particular, the progress in molecular biology enable us to create and/or discover new high-value biomolecules, with the rapid progress in genetic engineering and the development of many new recombinant DNA technique. Metabolic engineering has been defined as "purposeful modification of intermediary metabolism using recombinant DNA techniques"[2]. Originally, the term metabolic engineering has been defined as "Improvement of cellular activities by manipulation of enzymatic transport and regulatory functions of the cell with the use of recombinant DNA technology" [3]. In a broader sense, metabolic engineering can be viewed as the design of biochemical reaction networks to accomplish a certain objective. Typically, the objective is either to increase the rate of a desired product or to reduce the rate of undesired side-products [4], or to decompose the toxic or undesired substances. Of central importance to this field is the notion of cellular metabolism as a network. In other words, an enhanced perspective of metabolism and cellular function can be obtained by considering the participating reactions in their entirety, rather than on an individual basis [4]. This research field is, therefore, multidisciplinary, drawing on information and techniques from biochemistry, genetics, molecular biology, cell physiology, chemistry, chemical engineering, systems science, and computer science.

Although the term metabolic engineering has been defined as above, its definition is by no means clear-cut, and overlaps with the related terms such as physiological engineering [5], pathway engineering, in vitro evolution, direct evolution, molecular breeding, cellular engineering as listed in Cameron and Tong [2].

The intellectual framework and the potential application of metabolic engineering have been reviewed [3,6], and yet another reviews have been made to recognize the importance of metabolic engineering [2,5,7-9]. Several books have also been published recently [10,11].

In the present article, an overview on metabolic systems engineering approach is made for the efficient fermentation based on our recent research results. It should be noted that it is getting more important to investigate gene and protein expressions in order to uncover the metabolic regulation mechanism.

Metabolic Signal Flow Diagram Approach

One of the objectives of metabolic engineering is to understand how culture environment affects metabolic pathways, since it eventually guides the development of gene engineering and cultivation techniques in a direction toward increased yield or productivity. From the engineering point of view, it may not be necessary to understand every steps of a metabolic reaction or the activity of every enzymes in a cell, since the important thing is to direct metabolic flux to the desired branch through which target products are produced.

From such a viewpoint, a graph-theoretic approach has been developed by Endo and his coworkers [12,13]. With this approach, it may be possible to determine the flow direction of each metabolic pathway for elucidation of the relationship between culture conditions, cell growth, and product formation. In principle, the metabolic networks can be expressed in terms of a directed signal flow diagram. Therefore, the enzyme reaction of conversion of metabolite A to metabolite B can be considered as the transformation of signal A to signal B. If we consider the metabolic network of a cell as a system, there are several inputs and outputs through the boundary of the system with the environment. Then those input-output relationships may be expressed in terms of the metabolic transfer coefficients. The resulting complex signal flow diagram can be simplified through use of the equivalent transformation of graph theory. It should be noted that the input-output relationship may be identified from the time-series data. Then we may be able to find some relationships on the metabolic transfer coefficients, and thus we may be able to estimate the activities of certain pathway networks [12,14].

In relation to this approach, an interesting network analysis method has been proposed for formulation of a metabolic model for *Escherichia coli* by several researchers [15-17]. In their works, a metabolic objective provides a physiological rationale for acetate production, which is based on mechanistic details considered as constraints on the reaction network.

In the same manner, we assumed that a microorganism has as its objective maximization of ATP production and derived the expression of the metabolic transfer coefficients [18]. Such expression enables us to understand how the important branch points are regulated based on the time-series input and output data. This method was extended for use in on-line estimation [19]. For the on-line analysis, Shimizu et al.[20] defined an error vector based on stoichiometric equations, and they attempted to identify the unknown metabolites based on the magnitude of such an error vector due to unbalance of material.

The next question is whether we can determine in detail every enzyme reactions in the metabolism occurring inside the viable cell. This is the next topic to be discussed.

Metabolic Flux Analysis

Quantification of metabolic fluxes is an important analysis technique of metabolic engineering. A powerful technique for calculation of the fluxes through various pathways is the so-called metabolic flux analysis (MFA), where the intracellular fluxes are calculated using a stoichiometric model for all the major intracellular reactions and by applying mass balances around the intracellular metabolites. As inputs to the calculations, a set of measured fluxes, typically the specific uptake rate of substrate and the specific secretion rate of metabolites etc. are provided [21-23]. Metabolic flux distribution can then be estimated using the following stoichiometric equation:

$$Ar = q \qquad (1)$$

where A is an m x n matrix of stochiometric coefficients. r is an n-dimensional flux vector, and q is an m-dimensional vector of the specific substrate consumption rate and the specific metabolite excretion rates. The weighted least square solution to Eq.(1) is then obtained for the over-determined system as

$$r = (A^T \, \Phi^{-1} A)^{-1} A^T \, \Phi^{-1} q \qquad (2)$$

provided that A is of full rank, where Φ is the measurement noise variance-covariance matrix of the measurement vector q. Several computer programs for calculating r have been developed by several researchers [24,25].

We made flux analysis for efficient production of poly β-hydroxybutyrate (PHB) using *Ralstonia eutropha*. PHB is a homopolymer of 3-hydroxybutyrate and is most widespread and best-characterized member of poly-3-hydroxyalkanoates (PHAs). We conducted several fed-batch experiments for several substrates, and computed the flux distribution for the subdivided growth, transient, and PHB production phases.

Note that NH_3 concentration was relatively high during cell growth phase while it was low during the later PHB production phase. Because of this, the ammonium ion was dominantly assimilated into the cell through the reaction from α-ketoglutarate (α-KG) to glutamic acid (GLUT) during cell growth phase. It should also be noted that NADPH generated via isocitrate dehydrogenase (ICD) was mainly consumed in glutamic acid synthesis pathway during this phase. We estimated how much and where NADPH was produced and consumed in each cultivation phase based on the flux distribution obtained. The result clearly indicates that the block in the amino acid synthetic pathway due to low level of NH_3 in the later phase results in the overproduction of NADPH through ICD and accelerates the biosynthesis of PHB since NADPH-dependent acetoacetyl-CoA reductase can provide a sink for excess reducing equivalents. This research result demonstrates that MFA may be useful in disclosing the metabolic regulation mechanism to some extent [26]. We also made such MFA for other fermentation systems such as photosynthetic microorganisms [27,28].

Gulik and Heijnen [29] made 99-demensional metabolic flux analysis of aerobic growth of *S. cerevisiae* on glucose/ethanol mixtures and predicted five different metabolic flux regimes upon transition from 100 % glucose to 100 % ethanol. Pramanik and Keasling [30] developed a stoichiometric model for *E. coli* which incorporates 153 reversible and 147 irreversible reactions and 289 metabolites from metabolic data bases for the biosynthesis of the macromolecular precursors, coenzymes, and prosthetic groups necessary for synthesis of all cellular macromolecules. For such a system, the number of reactions is greater than the number of metabolites. Because multiple solutions exist, linear optimization was used to determine the fluxes. Some of the

12

objective functions that were used included minimization or maximization of ATP usage, substrate uptake, growth rate, and product synthesis.

Takiguchi et al.[31] applied the metabolic pathway model to estimate the physiological state of the cells, that is, the growth and production activity, and the flux distribution of metabolites for lysine fermentation using *Corynebacterium glutamicum* from on-line measurable data.

Metabolic Flux Analysis Using Isotopomer Distributions

The metabolic flux analysis based on the metabolic flux distributions computed by the above method may be useful to find the metabolic regulation mechanism for performance improvement of fermentation etc. However, the application of the above method is limited to relatively simple case. In more complex metabolic network systems, where product formation occurs simultaneously, the system involves cyclic pathways such as TCA cycle etc. where intermediates reenter the cyclic pathway, or the large number of branching points exist in the metabolic network, a detailed analysis cannot be done with the above method. In such cases, the application of metabolite balancing either requires another sets of reactions or forced to be lumped together [21].

The modeling of isotope distributions may be used to evaluate intracellular fluxes in more detail and to overcome the shortcomings of the above conventional MFA method. Zupke and Stephanopoulos [32] introduced atom mapping matrices (AMMs) to formulate such a problem, where they describe the transfer of atoms from reactants to products. For each reaction, there will be mapping matrices for every reactant-product pair.

The set of equations describing isotope distributions in a metabolic network using AMMs can be solved iteratively via computer. This requires that the specific activities of the substrate carbon atoms be initialized (to values between 0 and 1 for fractional enrichment) and a consistent set of flux value be provided. Then, each of the steady-state equations is solved sequentially for the activities of the output metabolites, and the process is repeated until convergence is achieved.

The vector representation of metabolite atom specific activities enables the equations governing the isotopic steady-state to be written directly from mass balance equations. The mass balances in turn follow directly from the set of stoichiometric reactions that forms the biochemical network. The derivation of AMMs is performed separately and is independent of the biochemical network [32].

It should be noted that the labeling patterns of the intracellular metabolites are difficult to measure due to the small pool sizes of these metabolites. However, since the amino acids reflect the labeling patterns of a number of important central metabolites through their precursors from the central metabolism, and relatively abundant, the labeling patterns of amino acids have been used for elucidation of labeling patterns in the central metabolism [33]. Thus ^{13}C-NMR and GC-MS have been employed to identify indirectly the labeling patterns of the intracellular metabolites. In the isotope labeling experiments, the mixture of 90% of unlabeled or naturally labeled carbon source such as glucose and 10% of uniformly labeled carbon source [U-^{13}C] and/or the first carbon labeled carbon source [1-^{13}C] is employed. In the analysis of carbon labeling patterns of amino acids, it is also necessary to make a correction for the contribution of labeling arising naturally labeled species of ^{15}N, ^{17}O, ^{18}O, ^{13}C, ^{2}H etc. This correction may be carried out using matrix-based methods as given by Lee et al. [34,35] and Wittman and Heinzle [36].

Marx et al. [23] combined the information of ^{1}H-detected ^{13}C NMR spectroscopy to follow individual carbons with carbon balances for cultivation of lysine-producing strain of *C. glutamicum*. The result shows that the flux through pentose phosphate pathway is 66.4 % (relative to the glucose input flux of 1.49 mmol/g dry weight h), that the entry into TCA cycle 62.2 %, and the contribution of the succinylase pathway of lysine synthesis 13.7 %.

For the systems having cyclic pathways, Klapa et al. [37] presented a mathematical model to analyze isotopomer distributions of TCA cycle intermediates following the administration of ^{13}C (or ^{14}C) labled substances. Such theory provides the basis to analyze ^{13}C NMR spectra and molecular weight distributions of metabolites. This method was applied to the analysis of several cases of biological significance [38].

Wiechert and his co-workers developed a model for positional carbon labeling systems, which was further extended to general isotopomer labeling systems with statistical treatment [39-41]. Mollney et al. [42] compared the different measurement techniques such as ^{1}H NMR, ^{13}C NMR and mass spectrometry (MS) as well as two-dimensional ^{1}H-^{13}C NMR techniques to characterize in more detail with respect to the formulation of measurement equations. Based on these measurement equations, a statistically optimal flux estimator was established. Having implemented these tools, different kinds of labling experiments were compared using statistical quality measures.

Although many applications have been reported for the flux analysis using NMR, GC-MS is also quite useful and has several advantages over NMR such that GC-MS require relatively small samples as compared with NMR and GC-MS gives rapid analysis etc. In GC-MS, the compounds are separated by the gas chromatography, and the mass spectrometry step analyzes the labeling patterns of the compounds as they elute. The mass spectrum of a compound usually contains ions that are produced by fragmentation of the molecular ion (i.e., the ionized intact molecule). These fragments contain different subsets of the original carbon skelton, and the mass isotopomer distributions of these fragments contain information that can, in addition to the information from the molecular ion, be used for analyzing the labeling pattern of the metabolites [43]. Christiensen and Nielsen [44] quantified the intracellular fluxes of Penicillium chrisogenum using GC-MS. They found that glycine was synthesized not only by serine hydroxymethyltransferase, but also by threonine aldorase. The formation of cytosolic acetyl-CoA was also found to be synthesized both via the citrate lyase-catalysed reaction and by degradation of the penicillin side-chain precursor, phenoxyacetic acid.

We recently used both ^{1}H-^{13}C 2D NMR and GC-MS to quantify the flux distributions of cyanobacteria at different culture conditions such as autotrophic, mixotrophic, and heterotrophic conditions, and obtained some insight into the metabolic regulation with respect to culture environment [45].

Synthesis of Metabolic Pathways and Optimization of Metabolic Flux Distribution

The synthesis of metabolic pathways involves the construction of pathways, i.e., sets of enzyme-catalyzed bioreactions whose stoichiometry is given, to meet certain specifications. Systematic synthesis of pathways that satisfy a set of specifications is relevant in the early steps of the conception and design of a bioprocess, where a pathway must be chosen for the production of the desired product. The first effort for systematic synthesis of metabolic pathways was made by Seressiotis and Bailey [46]. They presented an approach for synthesizing pathways that start from a given substrate and produce a target product with calculation of yield etc.

Mavrovouniotis et al. [47] developed a computer-aided synthesis method of metabolic pathways, where the algorithm satisfies each stoichiometric constraint by recursively transforming a base-set of pathways. They applied the algorithm to the problem of lysine synthesis from glucose and ammonia and discovered that the yield of lysine over glucose cannot exceed 67 % in the absence of enzymatic recovery of carbon dioxide.

Another approach is to find the optimal flux distribution based on known metabolic network structure and the objective function such as yield etc. We considered this problem for PHB production using recombinant E.coli. The genes involved in PHB biosynthesis in R.eutropha were cloned in E. coli, and were sequence l and characterized in detail by several researchers. Three genes form an operon in the order of phb-C-A-B, coding for PHB synthase, β-ketothiolase, and NADPH-dependent acetoacetyl-CoA reductase, respectively. Recombinant E.coli provides several advantages over R. eutropha, such as the ability to utilize several inexpensive carbon sources, easy recovery of PHB, and no degradation of PHB once synthesized. However, PHB yield is not high enough depending on the culture media. The regulatory mechanism for PHB synthesis in recombinant E. coli is markedly different from that of R.eutropha since the former can efficiently synthesize PHB during growth without any nutrient limitation such as NH_3 limitation. We addressed the problem of how much PHB can be theoretically produced by recombinant *E. coli* considering the constraints imposed by stoichiometry. The steady-state flux balance was formulated as a linear programming problem in which the production of PHB was maximized subject to the stoichiometric constraints.

The predicted flux distributions indicate that in order to achieve the maximum PHB yield, about half of the carbon flow should be metabolized via the pentose-phosphate (PP) pathway and the flux to TCA cycle should be shut down. This is easy to understand since there are two pathways signficantly affect the availability of two substrates for PHB synthesis, NADPH and acetyl-CoA. The predicted flux from glucose 6-phosphate to the PP pathway was, however, 46 % which is much higher than the normal value of 28 % [48]. If the flux to PP pathway was restricted to 28 % of glucose carbon, the PHB yield decreased from 0.92 to 0.82. Based on our analysis, the metabolic pathway that metabolizes glucose through part of the PP pathway and the Entner-Doudoroff (ED) pathway may be the best pathway for a metabolic variant of *E. coli* to overproduce PHB [49]. This kind of analysis gives the upper bound for the yield etc. and may give the motivation of whether further effort should be made or not.

Modification of Metabolic Pathways

(1) Gene-engineering technique
Optimization of the metabolic flux distribution of recombinant *E. coli* for PHB production illustrated the importance of the availability of acetyl-CoA and NADPH for achieving the maximum yield of PHB. In order to examine whether the increased availability of the above substances can enhance PHB synthesis in recombinant *E. coli*, several *E. coli* K12 derivatives, namely, HMS174, TA3516 (*pta⁻/ack⁻*), and DF11 (*pgi⁻*), were transformed with a plasmid which contains the native phb operon. The fermentation characteristics of these recombinant strains were studied and compared. We then examined the effects of intracellular acetyl-CoA accumulation, which may promote PHB synthesis *in vivo*, by perturbations induced from attenuation of acetate kinase and phosphotransacetylase and by cultivation of *E. coli* HMS174 using gluconate as a substrate; it can convert gluconate to acetyl-CoA at higher rate. The effect of intracellurar accumulation of NADPH were investigated by introducing a perturbation induced from attenuation of phosphoglucose isomerase, which redirects the carbon flow to the PP pathway. Results indicate that intracellular buildup of acetyl-CoA may not be able to promote PHB synthesis *in vivo*. On the other hand, since the biosynthesis of PHB in the *pgi⁻* mutant strain can utilize the NADPH overproduced through the PP pathway, the growth of the *pgi⁻*

mutant on glucose was recovered, indicating that the overproduction of NADPH might be able to enhance PHB synthesis [49].

Consider lactic acid fermentation as another example of modifying metabolic pathways by gene-engineering technique. In lactic acid fermentation, lactic acid production increasingly inhibits cellular metabolic activities. Several processes have, therefore, been developed to remove the lactic acid *in situ*. Among them, extractive fermentation has recently been paid most attention. The problem with this process, however, is that only undissociated lactate is extracted. Since in lactic acid fermentation, the culture pH is generally maintained at around 6-7, the direct application of extractive fermentation yields very low productivity due to large fraction of dissociated lactate at such pH values.

We, therefore, considered to modify the metabolism of *Saccharomyces cerevisiae* by expressing the lactate dehydrogenase (LDH) gene to produce a relevant amount of lactic acid at low pH. The plasmid pADNS which contains the ADH1 promoter was used as a host vector, and a heterologous gene region, cDNA-LDH-A (encoding bovine lactate dehydrogenase) digested from plasmid pLDH12 was digested and ligated into the two host vectors. The resultant plasmids were transformed into *S. cerevisiae* DS37. Using this recombinant *S. cerevisiae* strain, several fermentations were conducted at several pH values (4.5-3.5). Since the recombinant *S. cerevisiae* produced a considerable amount of ethanol as well as lactate (about 10 g/l), we then disrupted several pyruvate decarboxylase (PDC) genes to suppress the ethanol formation. Among the *PDC* genes, *PDC1*, *PDC5*, and *PDC6*, *PDC1* showed the greatest effect on the cell growth and ethanol production. The plasmid which contains the *LDH-A* structure gene was then transformed into the mutant strain lacking the *PDC1* gene. Cultivation of this strain improved the lactate yield from glucose while suppressing ethanol formation [50].

Those results indicate that the metabolic pathway modification can be made by gene-engineering technique locally, but it is not an easy task to modify metabolic flux distribution to the desired one.

(2) Control of culture environment

Another means of modifying the metabolic flux distribution is to manipulate the culture environment. We considered this problem for pyruvate fermentation. There is an increasing demand for pyruvic acid since it is an important raw material for the production of many amino acids such as tryptophan and tyrosine, and for the synthesis of many drugs and agrochemicals. Only a small amount of pyruvate is, however, secreted into the culture broth from wild-type microorganisms. Some auxotrophic strains, such as thiamine auxotrophs or lipoic acid auxotrophs have been utilized for pyruvate overproduction.

We investigated the metabolism and fermentation characteristics of *Torulopsis glabrata* IFO 0005, a vitamin-auxotrophic pyruvate-producing yeast screened by Yonehara and Miyata [51]. The strain used is auxotrophic for four vitamins such as thiamine hydrochloride, nicotinic acid (NA), biotin and pyridoxine hydrochloride. Since dissolved oxygen (DO) is an important environmental factor which affects the tricarboxylic acid (TCA) cycle activity, and thiamin hydrochrolide plays a key role in the strain [52], attention was focused on analyzing the effects of these culture conditions on the cellular metabolism of this strain. Several batch experiments were conducted at different DO concentrations and metabolic flux distributions were computed for each experiment for DO=40 %. The point in this research is to control the specific pathway pinpoint by manipulating culture environment and by addition of vitamins [53-55].

Metabolic Control Analysis

The concept of metabolic flux analysis is useful for quantification of flux distribution, but it does not allow for evaluation of how the fluxes are controlled. One of the most important aspects of metabolic engineering may be control of flux. The main objective is to find rate-limiting step and bottleneck enzyme. Delgado and Liao [56] proposed an idea of inverse flux analysis which allows the prediction of the flux distribution when some of the manipulable fluxes were perturbed. The application of this method to *E. coli* suggested that the increase in the flux of the anaplerotic pathways, indicating the reactions catalyzed by phosphoenol pyruvate carboxylase and the glyoxylate bypass will decrease acetate production while increasing the growth yield.

The concept of metabolic control analysis (MCA) was developed by Kacser and Burns [57] and Heinrich and Rapoport [58]. Several reviews have been made with some extensions [7,59,60]. In MCA, several kinds of coefficients play important roles. Elasticity coefficient (EC) is defined as

$$\varepsilon_k^i \equiv \frac{dv_i}{dx_k} \frac{x_k}{v_i} = \frac{d \ln v_i}{d \ln x_k} \qquad (3)$$

where v_i is the i-th reaction rate and x_k a k-th variable that modifies the rate. This coefficient is modulated by metabolites, which is of practical concern in enzyme regulation. If parameter p_k is used instead of x_k, it may be called as π-elasticity [61] instead of ε-elasticity.

Flux control coefficient (FCC) is defined [57] as

$$C_i^{Jk} \equiv \frac{dJ_k}{de_i} \frac{e_i}{J_k} = \frac{d \ln J_k}{d \ln e_i} \qquad (4)$$

where FCCs express the fractional change in the steady state flux through the pathway (J_k) that results from an infinitesimal change in the activity of enzymes (or reaction rates). It should be noted that whereas the elasticity coefficients are properties of the individual enzymes, the FCCs are properties of the system.

The FCCs are, therefore, not fixed but change with the environmental conditions [5].

Concentration control coefficient (CCC) is defined as

$$C_i^k \equiv \frac{dx_k}{de_i}\frac{e_i}{x_k} = \frac{d\ln x_k}{d\ln e_i} \quad (5)$$

We applied MCA for lysine fermentation. We studied how FCC changed as fermentation proceeds. It was found that the bottleneck enzyme changes: aspartokinase during lysine formation phase while permease at the late stage fermentation. We could increase the lysine production rate by increasing those enzyme activities by constructing gene-engineered *C. glutamucum* [62,63]. Nielsen and Jorgensen [64] have made the similar analysis for penicillin biosynthesis using *Penicillium chrisogenum*.

Although MCA approach has been enthusiastically employed by many researchers [65], the application to practical experimental systems is quite limited. Liao and Delgado [7] pinpointed the gaps still remaining between mathematical treatment and experimental implementations.

Stephanopoulos and Simpson [66] proposed a means for calculating group control coefficients as measures of the control exercised by groups of reactions on the overall network fluxes and intracellular metabolites pools. The concepts of this method were illustrated through the simulation of a case study involving aromatic amino-acid biosynthetic pathway. It was further demonstrated that the optimal strategy for the effective increase of network fluxes was through the coordinated amplification of a small number of steps in order to maintain maximum throughput while ensuring an uninterrupted supply of intermediate metabolites. Simpson et al [67] also proposed a method of determining group flux control coefficients based on many experimental data without using mathematical models.

Dynamic extension of metabolic control analysis has been made with dynamic model for the metabolic change on the order of minutes [68-70].

MCA is not a modeling framework, but it is a set of postulates that allows the systematic computation of network sensitivities to single perturbations in the environmental or network parameters. Thus its predictive capability is limited by the quality of the model employed. Since traditional kinetic models lack the description of the regulatory component, the sensitivity coefficients for MCA is suspect. The cybernetic framework developed by Ramkrishna and his co-workers may offer some advantage over conventional methodologies [71,72]. The framework hypothesizes that metabolic systems have evolved optimal goal oriented strategies as a result of evolutionary processes. The inclusion of a goal-oriented regulatory strategy gives the cybernetic description of a metabolic network, the key feature of regulatory responsiveness, an element that is missing from many other contemporary metabolic network analysis and modeling frameworks.

As stated above, MCA is based on information of the kinetics of individual reactions. The lack of *in vivo* kinetic information on the individual pathway reactions are limiting a widespread use of this concept. To overcome such a problem, there are yet alternative approaches based on thermodynamic considerations. The relevant thermodynamic variable for a thermodynamic feasibility analysis is the Gibb's free energy of a reaction (ΔG). A strict requirement for flow of carbon through cellular pathways is that ΔG is negative for all individual reactions. Mavrovouniotis [73,74] used this concept in order to develop a procedure that seeks the range of metabolic concentrations where all the reactions are feasible ($\Delta G < 0$). This procedure is based only on the knowledge of the value of $\Delta G^{0'}$, the standard Gibbs free energy and the concentration of cofactors. Pissara and Nielson [75] applied the thermodynamic feasibility algorithm of Mavrovouniotis [73,74] to the α-aminoadipic acid pathway in *P. chrysogenum*. Thereafter they considered the penicillin biosynthetic pathway, for which they calculated the standard Gibbs free energies using the concept of group contributions [73], and they used measurements of the pathway metabolite concentrations to calculate the ΔG values for the reactions during a fed-batch cultivation.

Design of Metabolic Regulatory Structures

It has recently been shown that genetic modification of metabolic control systems can significantly enhance the process performance. However, past modifications of enzyme properties in metabolic system were usually based on trial and error methods, which becomes ineffective when the system becomes optimization method is the utilization of kinetic model derived by the S-system complex. Since the recent genetic engineering techniques allow modifications of both gene-level (expression of genes) and protein-level (activity of enzymes) regulations, it is desired to develop a useful method for finding the optimal metabolic structure. However, the optimization techniques for metabolic system design are not well established now. One metabolic optimization method is to maximize the objective metabolic flux based on the stoichiometric information of a metabolic system. This approach may provide the optimal flux distributions for individual pathways of a metabolic network. However, it does not suggest the effects of the modification of metabolic regulatory structure, since no kinetics are considered in this method. Another conventionally used optimization method is the utilization of the kinetic model deried by the S-system formalism [76,77]. This approach does not also address the effect of the change in regulatory structure, although the optimum manipulation of external inputs can be provided [78-80]. The only way to find the optimal regulatory structures for a metabolic network may be to use the kinetic model obtained from enzymatic reaction mechanism. Since the mechanism-based kinetic expressions are usually nonlinear, and many data are required for the

20

optimization of nonlinear system, a (log)linearized model formulation has been developed by Hatzimanikatis et al. [81]. This type of (log)linearized model has been verified to approximate the original nonlinear model much better than the general linearized model [82]. By constructing a set of regulatory structures in which every metabolite is considered to be capable of regulating any enzyme in the system, and by considering every possible combination, the optimization can be carried out effectively for the nonlinear systems, and the optimal regulatory structures obtained can give the direction on how genetic engineering will be applied.

We attempted to find the optimal metabolic regulation structure by applying the mixed integer linear programming method to the log (linearized) model for the lysine synthetic pathway. The results indicate that more than 20% increase of internal lysine flux can be obtained when only the inhibitory regulation was allowed, and eight optimal structures with one regulatory loop were adopted. When regulation of enzyme activation was allowed, internal lysine flux can be increased by more than 70%. Changes of participating precursor and cofactor concentrations may not improve lysine flux significantly in this system [83].

Identification of Unknown Pathways

Historically, the isotopomer has been used to find the unknown pathways. However, the method itself is not systematic. It is expected that the metabolic synthesis methodology be extended for identifying unknown pathways. Here I just give our recent experiences for finding the new pathways in vitamin B_{12} fermentation.

When propionibacteria are grown under anaerobic condition, propionate, acetate, and CO_2 are produced through the randomizing pathway. Moreover, vitamin B_{12} is accumulated intracellularly under anaerobic condition. The accumulation of propionate causes strong inhibition on cell growth as well as vitamin B_{12} synthesis. In contrast, if cultivation was switched from anaerobic to aerobic, the propionate and vitamin B_{12} ceased to be produced while acetate continued to be produced. Noting the experimental fact that the propionate accumulated during anaerobic cultivation was decomposed when the cultivation was shifted to an aerobic condition, we considered to change the DO concentration periodically between 0 ppm and 1 ppm to keep the propionate concentration at low level and thus to show the improvement of vitamin B_{12} productivity [84].

Although the metabolic pathway of propionibacterium grown under anaerobic condition has been well investigated, its pathway under aerobic condition is not fully investigated yet. The randomizing pathway may function in a reversed direction in the presence of oxygen, through which the propionate is oxidized. The overall reaction of this pathway may be expressed as: Propionate→

pyruvate + 4H. Four electrons are released in this direction. In fact, the pyruvate concentration increased in the presence of oxygen. This hypothesis was supported by the analysis of isotope ^{13}C assignation. The $^{13}C1$ of propionate was first assigned to $^{13}C1$ of propionyl-CoA in reaction (Propionate + succinyl-CoA→ propionyl-CoA + succinate) catalyzed by propionyl-CoA transferase. It was, thereafter, assigned to C2 of pyruvate in reaction (Oxaloacetate + propionyl-CoA → pyruvate + methylmalonyl-CoA(a)) catalyzed by methylmalonyl-CoA: pyruvate carboxyl-transferase. It was further assigned to C1 of acetate in the oxidation of pyruvate. From the analysis, the direct oxidation of propionate to acetate could be excluded, although it may happen due to the presence of propionyl-CoA transferase. If we assume that CoA was transferred from acetyl-CoA to propionyl-CoA as shown in reaction (Propionate+acetyl-CoA→acetate+propionyl-CoA) under catabolization of propionyl-CoA transferase, the peak assigned to 2-^{13}C of pyruvate will not appear in the spectra of ^{13}C NMR. In fact, this peak appeared (data not shown). Thus it can be said that propionate is not directly oxidized to acetate [85].

Futeure Petspectives

Recent rapid progress in molecular biology unveiled the intricacies and mechanistic details of genetic information transfer and determined the structure of DNA and the nature of the gene code, establishing DNA as the source of heredity containing the blueprints form which organisms are built. These scientific revolutionary endeavors have rapidly spawned the development of new technologies and emerging fields of research based around genome sequencing efforts (i.e., functional genomics, structural genomics, proteomics and bioinfomatics) [86].

Once presented with the sequence of a genome, the first step is to identify the location and size of genes and their open reading frames (ORF's). DNA sequence data then need to be translated into functional information, both in terms of the biochemical function of individual genes, as well as their systematic role in the operation of multigenic functions.

Genomic technologies can be broadly divided into two groups, namely those that alter gene structure and deduce information from altered gene function and those that observe the behavior of intact genes [87]. The former methods involve the random or systematic alteration of genes across an organism's genome to obtain functional information. Current genome-altering technologies are distinguished by their strategies for mutagenizing and analyzing cell populations in a genome-wide manner.

For the latter, several methods have been developed to efficiently monitor the behavior of thousands of intact genes. DNA chips are tools that fractionate a heterogeneous DNA mixture into unknown components, while a complementary method, called SAGE (Serial Analysis of Gene Expression), has been important for identifying transcripts not predicted by sequence information alone [88]. The two commonly available DNA chips are oligonucleatide chips and DNA microarrays. A primary technical difference between oligonucleotide chips and DNA microarrays is the size of their DNA targets. DeRishi et al. [89] showed how the metabolic and genetic control of gene expression could be studied on a genome scale using DNA microarray technology. The temporal changes in genetic expression profiles that occur during the diauxic shift in *S. cerevisiae* were observed for every known expressed sequence tag in this genome.

One of the key issues in functional genomics is to relate linear sequence information to nonlinear cellular dynamics. Toward this end, significant scientific effort and resources are directed at mRNA expression monitoring methods and analysis, and at the same time the field of proteomics (the simultaneous analysis of total gene expression at the protein level) represents one of the premiere strategies for understanding the relationship between various expressed genes and gene products [90].

From annotated genomes, we can directly construct the stoichiometric matrix for the entire metabolic network of an organism, which allows us to determine metabolic pathways. The genome-specific stoichimetric matrix can be obtained for any recently sequenced and annotated genome, and it may be used to synthesize *in silico* organisms. With this thinking, Palsson and his co-workers [86,91] constructed an *in silico* strain of *E. coli* K-12 from annotated sequence data and from biochemical information. Using this *in silico* microorganism, one can study the relation between *E. coli* metabolic genotype and phenotype in the *in silico* knockout study [91]. At the heart of this persective is the study of the system as a whole rather than the detailed study of individual components and their direct interactions.

It should be noted that it is important to understand the regulation mechanism of gene and protein expressions as well as metabolic regulation. Recently, we investigated how culture environments affect those regulations for *Synechocystis* using RT-PCR and 2-dimensional electrophoresis as well as NMR and GC-MS [92]. We found that many genes are differentially regulated according to different mechanism, and the regulation of metabolic fluxes may be exerted at the transcriptional, post-transcriptional, translational, post-translational, and metabolic levels. Although at present, the transcriptomics, proteomics, and metabolic flux analysis allow high-throughput analysis of gene expression profiles, each of these techniques has its own advantages and limitation,, and only their integration may provide us with a detailed gene expression phenotype at each level and allow us to tackle the great complexity underlying biological processes.

Finally, it should be stated that a fusion of concepts from biological and nonbiological disciplines, including mathematics, computer science, physics, chemistry and engineering is required to address the theoretical and experimental challenges facing the field of genomics, and together promise great breakthroughs in our understanding and engineering cellular systems [87].

References

[1]Ryu, D.D.Y. and D-H. Nam, Recent Progress in Biomolecular Engineering, Biotechnol. Prog., **16**, 2-6(2000).

[2]Cameron, D.C. and I.-T. Tong, Cellular and Metabolic Engineering An Overview, Appl. Biotechnol., **38**, 105-140(1993).

[3]Bailey, J.E., Toward a science of metabolic engineering, Science, **252**, 1668-1675 (1991).

[4]Stephanopoulos, G., Metabolic Engineering," Curr. Opinion in Biotechnol., **5**, 196-200 (1994).

[5]Nielsen, J.,Metabolic Engineering: Techniques for Analysis of Targets for Genetic Manipulations, Biotechnol.Bioeng.,**58**, 125-132(1998).

[6]Stephanopoulos, G. and J.J. Vallino, Network rigidity and metabolic engineering in metabolite overproduction, Science, **252**, 1675-1681 (1991).

[7]Liao, J.C. and J. Delgado, Advances in Metabolic Control Analysis, Biotechnol. Prog., **9**, 221-233 (1993).

[8]Shimizu, K., An overview on metabolic systems engineering approach and its perspectives for efficient microbial fermentation, J.Chin. Inst. Chem.Eng.,**31**, 429-442(2000).

[9]Shimizu,K, Metabolic pathway engineering: Systems analysis methods and their applications, in Adv.in Appl.Biotechnol. (ed. By J.J.Zhong),ECUST in press, China (2000).

[10]Stephanopoulos,G., A.A.Aristidou, and J.Nielsen, Metabolic Engineering: Principles and Methodorogies, San Diego,CA, Academic Press(1999).

[11]Lee,S.Y. and T.Papoutsakis (eds.), Metabolic Engineering, NewYork, NY, Marcel Dekker (1999).

[12]Endo, I. and I. Inoue, Metabolic activities of yeast cells in batch culture, Kagaku Kogaku Roubunshu, **2**, 416-421 (1976).

[13]Inoue, I. and I. Endo, An analysis of yeast metabolism in continuous culture, Kagaku Kogaku, **37**, 69-75 (1973).

[14]Jin, S., K. Ye and K. Shimizu, Metabolic pathway analysis of recombinant *Saccharomyces cerevisiae* with a galactose-inducible promoter based on a signal flow modeling approach, J. Ferment. Bioeng., **80**, 541-551 (1995).

[15]Majewski, R.A. and M.M Damach, Simple constrained-optimization view of acetate overflow in E. coli, Biotechnol. Bioeng., **35**, 732-738 (1990).

[16]Ko, Y.F., W.E. Bently and W.A. Weigand, An integrated metabolic modeling approach to describe the energy efficiency of *E. coli* fermentations under oxygen- limited conditions : cellular energetics, carbon flux and acetate production, Biotechnol. Bioeng., **42**, 843-853 (1993).

24

[17]Ko, Y.F., W.E. Bently and W.A. Weigand, A Metabolic Model of Cellular Energetics and Carbon Flux during Aerobic *E.coli* Fermentation, Biotechnol. Bioeng., **43**, 847-855 (1994)

[18]Shi, H. and K. Shimizu, An Integrated Metabolic Pathway analysis Based on Metabolic Signal Flow Diagram and Cellular Energetic for *Saccharomyces cerevisiae*, J. Ferment. Bioeng., **83**, 275-280 (1997).

[19]Shi, H. and K. Shimizu, On-line metabolic pathway analysis based on Metabolic Signal Flow Diagram, Biotechnol. Bioeng., **58**, 141-148 (1998).

[20]Shimizu, H., K. Miura, S. Shioya and K. Suga, On-line State Recognition in Yeast Fed-Batch Culture Using Error Vectors, Biotechnol. Bioeng., **47**, 165-173(1995).

[21]Vallino, J.J. and G. Stephanopoulos, Metabolic Flux Distributions in *Corynebacterium glutamicum* During Growth and Lysine Overproduction, Biotechnol. Bioeng., **41**, 633-646 (1993).

[22]Zupke, C., Stephanopoulos, G., Intracellular flux analysis in hybridomas using mass balances and in vitro ^{13}C NMR, Biotech. Bioeng., **45**, 292-303(1995).

[23]Marx, A., A.A. de Graaf, W. Wiechert, L. Eggeling and H. Shohm, Determination of the Fluxes in the Central Metabolism of *Corynebacterium Glutamicum* by Nuclear Magnetic Resonance Spectroscopy Combined with Metabolite Balancing, Biotechnol. Bioeng., **49**, 111-129 (1996).

[24]Mavrovouniotis, M.L., Computer-Aided Design of Biochemical Pathways, PhD Thesis, MIT, Cambridge, MA (1989).

[25]Pissarra, P.N. and C.M. Henriksen, Fluxmap. A Visual Environment for Metabolic Flux Analysis of Biochemical Pathways, Preprint of the 7th Int. Conf. On Comp. Appl. In Biotechnol., Osaka, Japan, 339-344 (1998).

[26]Shi, H., M. Shiraishi and K. Shimizu, Metabolic Flux Analysis for Biosynthesis of Poly(β -Hydroxybutyric Acid) in *Alcaligenes eutrophus* from Various Carbon Sources, J. Ferment. Bioeng., **84**, 579-587 (1997).

[27]Hata, J., Q.Hua, C.Yang, K.Shimizu, M.Taya, Characterization of energy conversion based on metabolic flux analysis in mixotrophic liverwort cells, *Marchantia polymorpha*, Biochem.Eng.J., **6**, 65-74 (2000).

[28]Yang,C, Q.Hua, K.Shimizu, Energetics and carbon metabolism during growth of microalgal cells under photoautotrophic, mixotrophic and cyclic light-autotrophic/dark-heterotrophic conditions, Biochem. Eng. J., **6**, 87-102(2000)

[29]van Gulik, W.M. and J.J. Heijinen, A Metabolic Network Stochiometry Analysis of Microbial Growth and Product Formation, Biotechnol. Bioeng., **48**, 681-698 (1995).

[30]Pramanik, J. and J.D. Keasling, Stoichiometric Model of *Escherichia coli* Metabolism : Incorporation of Growth-Rate Dependent Biomass Composition and Mechanistic Energy Requirements, Biotechnol. Bioeng., **56**, 398-421 (1997).

[31]Takiguchi, N., H. Shimizu and S. Shioya, An On-line Physiological State Recognition System for the Lysine Fermentation Process Based on a Metabolic Reaction Model, Biotechnol .Bioeng., **55**, 170-181 (1997).

[32]Zupke, C. and G. Stephanopoulos, Modeling of Isotope Distributions and Intracellular Fluxes in Metabolic Networks Using Atom Mapping Matrices, Biotechnol. Prog., **10**, 489-498(1994).

[33]Szyperski ,T., Biosynthetically directed fractional ^{13}C-labeling of proteinogenic amino acids: An efficient analytical tool to investigate intermediary metabolism, Eur.J.Biochem., **232**, 433-448 (1995).

[34]Lee,W.-N. P., L.O. Byerley and E.A.Bergner, J.Edmond, Mass isotopomer analysis: Theoretical and practcal considerations, Biol. Mass Spectrometry, **20**,451-458 (1991).

[35]Lee,W.-N. P., E.A.Bergner, Z.K.Guo, Mass isotopomer patterns and precurser-product relationship, Biol. Mass Spectrometry, **21**,114-122 (1992).

[36]Wittmann, C., and E.Heinzle, Mass spectrometry for metabolic flux analysis, Biotech. Bioeng., **62**, 739-750 (1999).

[37]Klapa, M.I., S.M. Park, A.J. Sinskey and G. Stephanopoulos, Metabolite and Isotopomer Balancing in the Analysis of Metabolic Cycles: I. Theory, Biotechnol. Bioeng., **62**, 375-391 (1999).

[38]Park, S.M., M.I. Klapa, A.J. Sinskey and G. Stephanopoulos, Metabolite and isotopomer balancing in the analysis of metabolic cycles, Biotechnol. Bioeng., **62**, 392-401 (1999).

[39]Wiechert W. and A.A. de Graaf, Metabolic Networks: I. Modeling and Simulation of carbon Isotope Labelling Experiments, Biotechnol. Bioeng., **55**, 101-117 (1997).

[40]Wiechert W. C. Siefke, A.A. de Graaf and A. Marx, Metabolic Networks: II. Flux Estimation and Statistical Analysis, Biotechnol. Bioeng., **55**, 118-135 (1997).

[41]Wiechert W., M. Mollney, N. Isermann, M. Wurzel and A.A. de Graaf, Bidirectional Reaction Steps in Metabolic Networks: III. Explicit Solution and Analysis of Isotopomer Labeling Systems, Biotechnol. Bioeng., **66**, 69-85 (1999).

[42]Mollney, M., W. Wiechert, D. Kownatzki and A.A. de Graaf, Bidirectional Reaction Steps in Metabolic Networks: IV. Optimal Design of Isotopomer Labeling Experiments, Biotechnol. Bioeng., **60**, 86-103 (1999).

[43]Christensen,B., J.Nielsen, Isotopomer analysis using GC-MS, Metabolic Eng.,**1**,282-290(1999).

[44]Christensen,B., J.Nielsen, Metabolic network analysis of *P. chrisogenum* using ^{13}C-labeled glucose, Biotechnol. Bioeng., **68**, 652-659 (2000).

[45]Yang,C., Q.Hua, K.Shimizu, Quantitative analysis of intracellular metabolic fluxes using GC-MS and two-dimensional NMR spectroscopy, J.Biosci. Bioeng., **93**,78-87 (2002).

[46]Seressiotis, A. and J.E. Bailey, MPS : An Artificially Intelligent Software

System for the Analysis and Synthesis of Metabolic Pathways, Biotechnol. Bioeng., **31**, 587- 602 (1988).

[47]Mavrovouniotis, M.L., Greg. Stephanopoulos and George Stephanopoulos, Computer-Aided Synthesis of Biochemical Pathways, Biotechnol. Bioeng., **36**, 1119-1132 (1990).

[48]Holms, W.H., Flux Analysis and Control of the Central Metabolic Pathways in *Escherichia coli*, FEMS Microbial. Rev., **19**, 85-116(1996).

[49]Shi.H., J.Nikawa, K.Shimizu, Effect of modifying metabolic network on poly (3-hydroxybutyrate) biosynthesis in recobminant *Escherichia coli*., J. Biosci. Bioeng., **87**, 666-677 (1999).

[50]Adachi,E., M.Torigoe, M.Sugiyama, J.Nikawa, K.Shimizu, Modification of metabolic pathways of *S.cerevisiae* by the expression of lactate dehydrogenase and deletion of pyruvate decalboxylase genes for the lactic acid fermentation at low pH value, J.Ferment. Bioeng.,**86** (3) 284-289 (1998).

[51]Yonehara, T. and R. Miyata, Fermentation Production of Pyruvate from Glucose by *Torulopsis glabrata*, J. Ferment. Bioeng., **78**, 155-159 (1994).

[52]Miyata, R. and T.Yonehara, Improvement of Fermentative Production of Pyruvate from Glucose by *Torulopsis glabrata* IFO 0005, J. Ferment. Bioeng., **82**, 475-479 (1996).

[53]Hua, Q., C.Yang, K.Shimizu, Metabolic flux analysis for efficient pyruvate fermentation using vitamin-auxotrophic yeast of T.glabrata, J. Biosci. Bioeng.,.**87**, 206-213 (1999).

[54]Hua,Q., K.Shimizu, Effect of dissolved oxygen concentration on the intracellular flux distribution for pyruvate fermentation, J.Biotechnol.,**68**,135-147 (1999).

[55]Hua, Q., C.Yang, K.Shimizu,, Effect of glucose, vitamins, and DO concentrations on the pyruvate fermentation using T. glabrata IFO 0005 with metabolic flux analysis, Biotech.Prog., in press (2001).

[56]Delgado, J. and J.C. Liao, Inverse Flux Analysis for Reduction of Acetate Excretion in Escherichia coli, Biotechnol. Prog., **13**, 361-367 (1997).

[57]Kacser, H. and J.A.Burn, The Control of Flux, Symp. Soc. Exp. Biol., **27**, 65-104(1973).

[58]Heinrich, R., T.A.Rapoport, A Linear Steady-state Treatment of Enzymatic Chains. General Properties, Control and Effector Strength, Eur. J. , Biochem., **42**, 89- 95(1974).

[59]Liao, J.C. and J. Delgado, Flux Calculation Using Metabolic Control Constraints, Biotechnol. Prog., **14**, 554-560 (1998).

[60]Ehlde, M. and G. Zacchi, A general formalism for Metabolic Control Analysis, Chem. Eng. Sci., **52**, 2599-2606 (1997).

[61]Kacser, H., H.Sauro,and L. Acerenza, Enzyme-Enzyme Interactions and Control Analysis. 1. The case of Non-Additivity: Monomer-Oligomer Associations, Eur. J. Biochem., **187**, 481-491 (1990).

[62]Yang, C., Q. Hua and K. Shimizu, Development of a Kinetic Model for L-Lysine Biosynthesis in *Corynebacterium glutamicum* and Its Application to Metabolic Control Analysis, J. Biosci. Bioeng., **80**, 393-403 (1999).

[63]Hua,Q,C.Yang, K.Shimizu, Metabolic control analysis for lysine synthesis using *Corynebacterium glutamicum* and experimental verification, J.Biosci.Bioeng., **90**,184-192 (2000).

[64]Nielsen, J. and H.S. Jorgensen, Metabolic Control Analysis of the Penicillin Biosynthetic Pathway in a High-Yielding Strain of Penicillium chrisogenum, Biotechnol. Prog., **11**, 299-305 (1995).

[65]Fell, D., Metabolic Control Analysis : A Survey of Its Theoretical and Experimental Development , Biochem. J., **286**, 313-330(1992).

[66]Stephanopoulos, G. and T.W. Simpson, Flux Amplification in Complex Metabolic Networks, Chem. Eng. Sci., **52**, 2607-2627(1997).

[67]Simpson, T.W., H. Shimizu and G. Stephanopoulos, Experimental Determination of Group Flux Control Coefficients in Metabolic Networks, Biotechnol. Bioeng., **58**, 681-698 (1998).

[68]Mauch, K., S. Arnold and M.Reuss, Dynamic Sensitivity Analysis for Metabolic Systems, Chem. Eng. Sci., **52**, 2589-2598 (1997).

[69]Theobald, U., W. Mailinger, M. Baltes, M. Rizzi and M. Reuss, In Vivo Analysis of Metabolic Dynamics *in Saccharomyces cerevisiae* : I. Experimental Observations, Biotechnol. Bioeng., **55**, 305-316 (1997).

[70]Rizzi, M., M. Baltes, V. Theobald and M. Reuss, In Vivo Analysis of Metabolic Dynamics in Saccharomyces cerevisiae : II. Mathematical Model, Biotechnol. Bioeng., **55**, 592-608 (1997).

[71]Varner, J. and D. Ramkrishna, Metabolic Engineering from a Cybernetic Perspective. 1. Theoretical Preliminaries, Biotechnol. Prog., **15**, 407-425 (1999).

[72]Varner, J. and D. Ramkrishna, Metabolic Engineering from a Cybernetic Perspective. 2. Qualitative Investigation of Nodal Architechtures and Their Response to Genetic Perturbation, Biotechnol. Prog., **15**, 426-438 (1999).

[73]Mavrovouniotis, M.L., Group Contributions for Estimating Standard Gibbs Energies of Formation of Biochemical Compounds in Aqueous Solutions, Biotechnol. Bioeng., **36**, 1070-1082 (1990).

[74]Mavrovouniotis, M.L., Identification of Localized and Distributed Bottlenecks in Metabolic Pathways, International Conf. On Intelligent System for Molecular Biology, Washington DC, (1993).

[75]Pissarra, P.N. and J. Nielsen, Thermodynamics of Metabolic Pathways for Penicillin Production : Analysis of Thermodynamic Feasibility and Free Energy Changes During Fed-Batch Cultivation, Biotechnol. Prog., **13**, 156-165 (1997).

[76]Savageau, M.A., Voit, E.O. and Invine, D.H. Biochemical systems theory and metabolic control theory: 1. Fundamental similarities and differences. Math. Biosci. **86**, 127-145 (1987).

28

[77]Savageau, M.A., Voit, E.O. and Invine, D.H. Biochemical systems theory and metabolic control theory: 2. The role of summation and connectivity relationships. Math. Biosci. **86**, 147-169 (1987).

[78]Hatzimanikatis, V., Floudas, C. A. and Bailey, J. E. Optimization of regulatory architectures in metabolic reaction networks. Biotechnol. Bioeng. **52**, 485-500 (1996b).

[79]Regan, L., Bogle, D. and Dunnill, P. Simulation and optimization of metabolic pathways. Comp. Chem. Eng. **17**, 627-637 (1993).

[80]Voit, E. O. Optimization in integrated biochemical systems. Biotechnol. Bioeng. **40**, 572-582 (1992).

[81]Hatzimanikatis, V., Floudas, C. A. and Bailey, J. E. Analysis and design of metabolic reaction networks via mixed-integer linear optimization. AIChE J., **42**, 1277-1292 (1996a).

[82]Hatzimanikatis, V. and Bailey, J. E. MCA has more to say. J. Theor. Biol. **182**, 233-242 (1996).

[83]Hua, Q., C.Yang, K.Shimizu, Design of metabolic regulatory structures for enhanced lysine synthesis flux using (log) linearized kinetic models, Biochem. Eng. J., **9**,49-57 (2001).

[84]Ye, K., M.Shijo,S.Jin,K.Shimizu, Efficient production of Vitamin B_{12} from propionic acid bacteria under periodic variation of dissolved oxygen concentration, J.Ferment.Bioeng.,**82**, 484-491(1996).

[85]Ye,K., M.Shijo, K.Miyano, K.Shimizu, Metabolic pathway of Propionibacterium growing with oxygen: enzymes, ^{13}C NMR analysis and its application for vitamin B_{12} production with periodic fermentation, Biotech. Prog., **15**, 201-201 (1999).

[86]Schilling, C.H., J.S. Edwards and B.O. Palsson, Toward Metaboolic Phenomics : Analysis of Genomic Data Using Flux Balances, Biotechnol. Prog., **15**, 288-295 (1999).

[87]Kao, C.M., Functional Genomic Technologies: Creating New Paradiums for Fundamental and Applied Biology, Biotechnol. Prog., **15**, 304-311 (1999).

[88]Velculescu, V.E., L. Zhang, W. Zhou, J. Vogelstein, M.A. Basrai, D.E.J. Bassett, P. Hieter, B. Vogelstein and K.W. Kinzler, Characterization of the yeast transcriptome, Cell, **80**, 243-251 (1997).

[89]DeRisi, J.L., V.R. Iyer and P.O. Brown, Exploring the Metabolic and Genetic Control of Gene Expression on a Genomic Scale, Science, **278**, 680-686 (1997).

[90]Hatzimanikatis, V., L.H. Choe and K.H. Lee, Proteomics: Theoretical and Experimental Considerations, Biotechnol. Prog., **15**, 312-318 (1999).

[91]Edwards, J.S. and B.O. Palsson, *Escherichia coli* K-12 in silico: Definition of its metabolic genotype and analysis of its capabilities, Submitted for publication (2000).

[92]Yang,C., Q.Hua, K.Shimizu, Integration of the information from gene expression and metabolic fluxes for the analysis of the regulatory mechanisms in *Synechocystis*, Appl. Microbiol. Bioeng., in press (2002).

Chapter 3

Engineering of *Escherichia coli* Central Metabolic Pathways for the Production of Succinic Acid

S. Y. Lee and S. H. Hong

Metabolic and Biomolecular Engineering National Research Laboratory,
Department of Chemical Engineering and BioProcess Engineering
Research Center, Korea Advanced Institute of Science and Technology,
373–1 Kusong-dong, Yusong-gu, Taejon 305–701, Korea

Succinic acid is an important chemical which is used in a wide
range of applications. Recently, fermentative production of
succinic acid by employing metabolically engineered bacteria
has been drawing much attention. *Escherichia coli*, the best
studied bacterium, produces only a small amount of succinic
acid. To enhance the succinic acid flux, fluxes of PEP
carboxylation and pyruvate carboxylation, which link C_3- and
C_4-branch, were amplified. By the overexpression of PEP
carboxylase gene, succinic acid productivity increased by 36%.
The malic enzyme which converts pyruvate to malic acid was
overexpressed, and metabolic flux analysis and metabolic
control analysis were carried out to have an insight into the
metabolism. Intracellular metabolic fluxes were optimized
according to the simulation results, and succinic acid
productivity could be increased by 700%, resulting in the 85%
of maximum theoretical yield.

© 2002 American Chemical Society

Succinic acid is a member of C_4-dicarboxylic acid family, and has been used in many industrial applications including a surfactant, an ion chelator, a food additive, and a supplement to pharmaceuticals, antibiotics and vitamins (Figure 1). Succinic acid can also be used as a precursor of several important chemicals such as 1,4-butanediol, tetrahydrofuran, γ-butyrolactone, and other C_4 chemicals (1). Especially, succinic acid is an intermediate of several green chemicals and materials. For example, the ester compound of succinic acid and 1,4-butanediol is used to make biodegradable polymer Bionelle by Showa Highpolymer Co., Ltd. (Tokyo, Japan). At present, succinic acid is manufactured by hydrogenation of maleic anhydride to succinic anhydride, followed by hydration to succinic acid. Only small amount of succinic acid is produced by microbial fermentation. Recently, much effort is being exerted for the production of succinic acid and its derivatives by microbial fermentation using renewable feedstocks (1).

Anaerobiospirillium succiniciproducens and *Actinobacillus succinogenes* have been shown to be the most efficient succinic acid producing strains (2, 3, 4). When, continuous cultures of *A. succiniciproducens* have been carried out, a volumetric productivity of 6.1 g succinic acid/L/h could be obtained (3, 5, 6). A rumen bacteria *A. succinogenes* 130Z also produces succinic acid to a high concentration from various carbon substrates such as glucose, fructose, lactose, maltose, mannitol, mannose, sucrose, xylose, and cellobiose (4, 7). The fluoroacetate-resistant mutant of *A. succinogenes* could produce 110 g/L succinic acid with the apparent yield of 120 mole succinic acid/100 mole glucose (7).

In these days, metabolic engineering has become a new paradigm for the more efficient production of desired bioproducts (8). Metabolic engineering can be defined as directed modification of cellular metabolism and properties through the introduction, deletion, and modification of metabolic pathways by using recombinant DNA and other molecular biological tools (8, 9). *E. coli* has been the workhorse for the production of recombinant proteins and various metabolites. These have recently been several studies on succinic acid production by metabolically engineered *E. coli*, which are reviewed in this chapter.

General strategies for succinic acid production by metabolically engineered *E. coli*

Anaerobic metabolic pathway of *E. coli* is presented in Figure 2. Under anaerobic condition, *E. coli* can produce various C_2, C_3, and C_4-compounds as final products. However, *E. coli* normally produces C_3-compound (lactic acid) or C_2-compound (acetic acid and ethanol) as major product. Only a small amount of C_4-compound (succinic acid) is produced (typically, 7.8% of total fermentation products) (10). Therefore, reorientation of carbon flux from the C_2-,

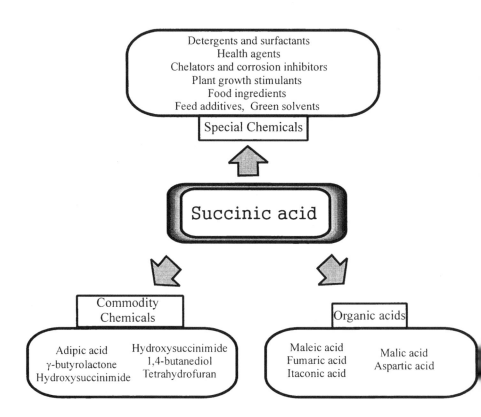

Figure 1. Various applications of succinic acid.

and C_3-branch to the C_4-branch is required for the enhanced production of succinic acid. The representative C_3-C_4 linkages present in *E. coli* are phosphoenolpyruvate (PEP) carboxylation and oxaloacetate decarboxylation. The malic enzyme coded by the *sfcA* gene also connects C_3- (pyruvate) and C_4-units (malic acid). However, under normal conditions, malic enzyme converts malic acid to pyruvate without reverse reaction (see below).

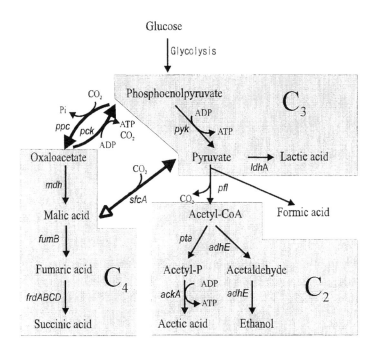

Figure 2. Anaerobic metabolic pathways in E. coli. *The thick arrows represent C_3-C_4 linkages. The open triangle direction does not normally operate in* E. coli. *Enzymes encoded by the genes shown are: ppc, PEP carboxylase; pck, PEP carboxykinase; mdh, Malate dehydrogenase; pyk, Pyruvate kinase; ldhA, Lactate dehydrogenase; pfl, Pyruvate formate-lyase; pta, Phosphotransacetylase; ackA, Acetate kinase; adhE, Alcohol/acetaldehyde dehydrogenase; sfcA, Malic enzyme; fumB, Fumarase B; frdABCD, Fumarate reductase.*

Amplification of PEP carboxylation flux

Millard et al. (*11*) reported the results of the amplification of PEP carboxylation reaction to direct more carbon flux toward C_4-branch. As presented in Figure 2, PEP carboxylase functions to direct PEP to oxaloactate. It is an important enzyme of the anepleoretic pathway, since it functions to supply additional oxaloacetate, which is consumed in biosynthetic pathways, to TCA cycle under aerobic condition. PEP carboxykinase plays an important role in gluconeogenesis. It catalyzes the phosphorylation of oxaloacetate to PEP with concomitant consumption of ATP. PEP carboxylase and PEP carboxykinase form a futile cycle, in which ATP is consumed without any product, when the PEP carboxykinase gene is overexpressed (*12*). Overexpression of PEP carboxylase gene resulted in a significant increase in the amount of succinic acid produced, while overexpression of PEP carboxykinase gene did not. When cells with amplified PEP carboxylase gene were cultivated in LB medium under anaerobic condition, the amount of succinic acid produced increased from 3.27 to 4.44 g/L (*11*).

The anaerobic PEP carboxykinase of *A. succiniciproducens* was also overexpressed in *E. coli* (*13*). It was supposed that the PEP carboxykinase of *A. succiniciproducens* is more suitable for the production of succinic acid since it produces ATP and conserves the free energy of PEP, while the PEP carboxylase of *E. coli* dissipates the free energy of PEP. The sequence of *A. succiniciproducens* PEP carboxykinase was found to be similar to those of all known ATP/ADP dependent PEP carboxykinases. Especially, the amino acid sequence of *A. succiniciproducens* PEP carboxykinase was 67.3% identical and 79.2% similar to that of *E. coli* PEP carboxykinase. Overexpression of the *A. succiniciproducens* PEP carboxykinase gene in *E. coli*, however, did not result in any increase in the amount of succinic acid produced (*13*).

Amplification of pyruvate carboxylation flux

The NAD^+-dependent malic enzyme was overexpressed to reorient pyruvate to C_4-branch and to enhance succinic acid production (*14, 15*). In normal *E. coli* metabolism, however, the majority of pyruvate is dissipated through C_2 and C_3- branches, and acetic and lactic acids are produced as major products under anaerobic condition. It is due to the higher affinities of lactate dehydrogenase and pyruvate formate-lyase to pyruvate compared to that of NAD^+-dependent malic enzyme. In addition, the K_m value of malic enzyme for pyruvate (16 mM) is much higher than that for malic acid (0.26 mM), which means that malic acid is not produced from pyruvate under normal cultivation condition. Therefore, a mutant *E. coli* strain NZN111, in which pyruvate-formate lyase and lactate dehydrogenase are partially blocked, was used. By double inactivation of pyruvate-formate lyase and lactate dehydrogenase genes, anaerobic pyruvate dissipation pathways are mostly blocked and the accumulated pyruvate is directed to C_4-branch by the malic enzyme (*14, 15*). Moreover, this strategy is superior in the energetic point of view. If two pathways of converting PEP to malic acid are compared, this latter pathway mediated by pyruvate kinase and malic enzyme is superior because ATP is produced when PEP is converted to pyruvate by pyruvate kinase, while the free energy of PEP is dissipated through the conversion of PEP to oxaloacetate (Figure 2). The NAD^+-dependent malic enzyme gene was overexpressed under the control of *trc* promoter in NZN111 to convert accumulated pyruvate to malic acid. Then, malic acid was further converted to succinic acid. At the end of cultivation in LB medium under anaerobic condition, 12.8 g/L succinic acid was produced and the apparent yield of succinic acid was 1.2 g succinic acid/g glucose. The malic enzyme of *Ascaris suum* was also overexpressed in *E. coli* NZN111 (*16*). The amount of succinic acid produced increased from 2.06 to 7.07 g/L, and succinic acid was produced as a major product.

In contrast with the results of flask culture, a considerable amount of malic acid, an intermediate metabolite in succinic acid producing pathway, was also produced in the fermentor scale studies. Therefore, metabolic flux analysis (MFA) was carried out to examine the change of metabolism in this recombinant NZN111 strain (*15*). By the use of MFA technique, the intracellular metabolic fluxes can be quantified by the measurement of extracellular metabolite concentrations in combination with the stoichiometry of intracellular reactions (*8, 17*). This system, however, could not be analyzed by conventional metabolic flux analysis techniques, since the intracellular accumulation of pyruvate was required for the conversion of pyruvate to malic acid (see above). Therefore, a new flux analysis method was proposed by introducing intracellular metabolite pools to mimic the intracellular accumulation of metabolites. From the results of MFA, it was found that conversion of malic acid to succinic acid became a new controlling step in this engineered pathway (*15*).

To identify the bottleneck, metabolic control analysis (MCA), which is a statistical modeling technique that can be used to understand the control of metabolic pathways and pathway regulations, was carried out. From the results of MCA, it was found that the succinic acid flux is highly sensitive to the activity of fumarate reductase. Redox balance is thought to be an important factor affecting the activity of fumarate reductase (*18*). According to the results of MCA, sorbitol which produces six moles of [H] during its conversion to PEP was examined as a carbon substrate to supply additional reducing power to the system (Figure 3). When NZN111 harboring pTrcML containing the *E. coli* NAD⁺-dependent malic enzyme was cultured in LB medium containing 20 g/L sorbitol in a 5 L fermentor, 10 g/L of succinic acid which is 7 times higher than that obtained with wild type *E. coli* strain was produced. The apparent yield of succinic acid was 1.1 g succinic acid/g sorbitol, which is 85% of the maximum theoretical yield (Hong and Lee, manuscript submitted).

Recently, a mutant strain of NZN111, AFP111 having a spontaneous *ptsG* gene (which encodes the glucose-specific permease of the phosphotransferase system) was isolated (*19, 20*). AFP111 produced 2 moles of succinic acid, 1 mole each of acetic acid and ethanol from 2 moles of glucose. When the disrupted *ptsG* gene was introduced to a *ldhA pfl* double mutant strain DC1327, the strain restored the glucose fermentation ability and produced succinic acid, acetic acid and ethanol at a molar ratio of 2:1:1, which is equivalent to those obtained with AFP111. From these results, it is obvious that inactivation of the *ptsG* gene redirect metabolic fluxes towards succinic acid production, even though the exact mechanism of this effect is not clear.

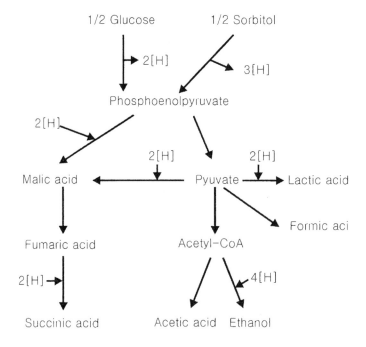

Figure 3. Redox balance in E. coli *NZN111 using glucose or sorbitol as carbon source. The formation and consumption of reducing power, [H], are shown.*

Conclusions

As reviewed above, several metabolic engineering strategies have been successfully employed for the enhanced production of succinic acid by *E. coli*. Through the amplification of the fluxes of PEP carboxylation and pyruvate carboxylation, succinic acid could be produced to a high concentration with a high yield and productivity in metabolically engineered *E. coli*. Furthermore, MFA and MCA allowed thorough understanding of metabolic pathways for designing better pathways (*15, 21*). As demonstrated in NZN111 fermentation studies, however, it is important to consider not only carbon flux but also the redox balance between the substrate and product for the production of primary metabolites.

Considering the limited nature of fossil fuels and pollution problems caused by petroleum industry, it is expected that biologically produced chemicals will eventually replace much of commodity chemicals currently based on petrochemicals. As further improvement of microbial strains is made by metabolic engineering, the cost effective production of succinic acid from renewable resources will be achieved.

Acknowledgement

Our work described in this paper was supported by the National Research Laboratory Program (2000-N-NL-01-C-237) of the Korean Ministry of Science and Technology (MOST).

References

1. Zeikus, J. G.; Jain, M. K.; Elankovan, P. *Appl. Microbiol. Biotechnol.* **1999**, *51*. 545-552.
2. Lee, P. C.; Lee, W. G.; Kwon, S.; Lee, S. Y.; Chang, H. N. *Enzyme. Microb. Technol.* **1999**, *24*, 549-554.
3. Lee, P. C.; Lee, W. G.; Kwon, S. H.; Lee, S. Y.; Chang, H. N. *Appl. Microbiol. Biotechnol.* **2000**, *54*, 23-27.
4. Van der Werf, M. J.; Guettler, M. V.; Jain, M. K.; Zeikus, J. G. *Arch. Microbiol.* **1997**, *167*, 332-342.
5. Lee, P. C.; Lee, W. G.; Lee, S. Y.; Chang, H. N. *Process Biochem.* **1999**, *35*, 49-55.

38

6. Lee, P. C.; Lee, W. G.; Lee, S. Y.; Chang, Y. K.; Chang, H. N. *Biotechnol. Bioprocess Eng.* **2000**, *5*, 379-381.
7. Guettler, M. V.; Rumler, D.; Jain, M. K. *Int. J. Syst. Bacteriol.* **1999**, *49*, 207-216.
8. Edwards, J. S.; Ramakrishna, R.; Schilling, C. H.; Palsson, B. O. In *Metabolic engineering*; Lee, S. Y.; Papoutsakis, E. T.; Ed.; Marcel Dekker: NY, 1999, pp 13-57.
9. Bailey, J. E. *Science* **1991**, *252*, 1668-1674.
10. Neijssel, O. M.; de Mattos, M. J.; Tempest, D. W. In Escherichia coli *and* Salmonella; Neidhardt, F. C.; Ed.; ASM press: WA, 1996, pp 1683-1692.
11. Millard, C. S.; Chao, Y.; Liao, J. C.; Donnelly, M. I. *Appl. Environ. Microbiol.* **1996**, *62*, 1808-1810.
12. Flores, C. L.; Gancedo, C. *FEBS Lett.* **1997**, *412(3)*, 531-534.
13. Laivenieks, M.; Vieille, C.; Zeikus, J. G. *Appl. Environ. Microbiol.* **1997**, *63*, 2273-2280.
14. Stols, L.; Donnelly, M. I. *Appl. Environ. Microbiol.* **1997**, *63*, 2695-2701.
15. Stols, L.; Kulkarni, G.; Harris, B. G.; Donnelly, M. I. *Appl. Biochem. Biotechnol.* **1997**, *63-65*, 153-158.
16. Hong, S. H., Lee, S. Y. *Biotechnol. Bioeng.* **2001**, *74(2)*, 89-95.
17. Nielsen, J.; Villadsen, J. In *Bioreaction engineering principles*; Nielsen, J.; Villadsen, J.; Ed.; Plenum press: NY, **1994**, pp 97-161.
18. Maklashina, E.; Cecchini, G. *Arch. Biochem. Biophys.* **1999**, *369(2)*, 223-232.
19. Donnelly, M. I.; Millard, C. S.; Clark, D. P.; Chen, M. J.; Rathke, J.W. *Appl. Biochem. Biotechnol.* **1998**, *70-72*, 187-198.
20. Chatterjee, R.; Millard, C. S.; Champion, K.; Clark, D. P.; Donnelly, M. I. *Appl. Environ. Microbiol.* **2001**, *67*, 148-154.
21. Hong, S. H.; Lee, S. Y. *J. Microbiol. Biotechnol.* **2000**, *10(4)*, 496-501.

Chapter 4

Metabolic Control Analysis in Glutamate Synthetic Pathway: Experimental Sensitivity Analysis at a Key Branch Point

Hiroshi Shimizu, Hisaya Tanaka, Akinori Nakato,
Keisuke Nagahisa, and Suteaki Shioya*

Department of Biotechnology, Graduate School of Engineering,
Osaka University, 2–1 Yamadaoka, Suita Osaka 565–0871 Japan

Experimental method for metabolic control analysis (MCA) was applied to investigation of a metabolic network of glutamate production by *Corynebacterium glutamicum*. Flux distribution at a key branched point of 2-oxoglutarate was investigated in detail. Enzyme activities of isocitrate dehydrogenase (ICDH), glutamate dehydrogenase (GDH), and 2-oxoglutarate dehydrogenase complex (ODHC) around the branched point were changed, using some genetically engineered strains and controlling environmental conditions. It was quantitatively found that the greatest impact on glutamate production around the branch point was obtained by attenuation of the ODHC activity.

© 2002 American Chemical Society

Introduction

Metabolic Engineering is a methodology of a target improvement for metabolite formation or cellular properties through the modification of specific biochemical reactions in the complicated metabolic networks (1, 2). To enhance the target product in bioprocesses, both genetic improvement of metabolic pathway and process operation strategies are very important. Recently wide varieties of methodologies in metabolic engineering have been developed: metabolic flux analysis based on the measurements of extracellular metabolites reaction rates (3) and/or ^{13}C isotope labeling and enrichment measurements in metabolites (4), analysis of control mechanism in the complicated metabolic networks (metabolic control analysis (MCA))(5, 6), and application of metabolic flux distribution analysis to operation and control of bioprocesses (7).

It has been well known that there are some triggering operations for glutamate production in *C. glutamicum*: depletion of biotin which is required for cell growth, addition of a detergent such as polyoxyethylene sorbitan monopalmitate (Tween 40), and addition of a lactum antibiotic such as penicillin. It was reported that decrease in activity of 2-oxoglutarate dehydrogenase complex (ODHC) and enhancement of glutamate production were observed after these triggering operation (8). The aim of this article is to determine quantitatively the degree of impact of changes in the enzyme activities around a key branch point, 2-oxoglutarate on a target flux of glutamate.

A metabolic reaction (MR) model was constructed for central carbon metabolism and glutamate synthetic pathways, and consistency of the model was statistically checked. Metabolic flux distribution was analyzed in detail by using the developed MR model. Especially, flux distribution at a key branched point of 2-oxoglutarate was investigated in detail. Enzyme activities of isocitrate dehydrogenase (ICDH), glutamate dehydrogenase (GDH), and 2-oxoglutarate dehydrogenase complex (ODHC) at the branched point were changed, using two genetically engineered strains and controlling environmental conditions. GDH and ICDH were enhanced in two transformants with each plasmid, which involved plasmids encoding homologous *gdh* and *icd* genes, respectively. ODHC activity was attenuated by biotin deficient condition.

The mole flux distributions in these strains were calculated by the metabolic reaction (MR) model, and the effect of the changes in the enzyme activities on the mole flux distributions were compared with each other. Sensitivity of enzyme activity changes on the glutamate production was evaluated by flux control coefficients (FCCs), which are defined by sensitivity of relative change in the steady state flux resulting from infinitesimal change in the activity of enzyme of the pathway (9). In this article, FCCs for the effect of enhancement of ICDH and GDH, and effect of attenuation of ODHC at the branch point, on the glutamate flux were quantitatively evaluated by the FCCs. The FCCs were experimentally determined by the method using large perturbation theory (5, 6) without precise kinetic information.

Materials and Methods

Microorganisms and Medium

Corynebacterium glutamicum AJ-1511 (ATCC13689) was used as a wild type strain of the glutamate producing microorganism. *C. glutamicum* AJ13678 and AJ13679 were genetically engineered strains which harbored plasmids with homologous *gdh* and *icd* genes, respectively. Strain, plasmid, relevant characteristics are summarized in Table I. Elemental composition analysis has shown the dry cell composition of AJ1511 strain to be $C_{4.17}H_{7.32}O_{1.92}N$ (ash content was 4.6%). It was assumed that the elemental compositions of the transformant strains of AJ13678 and AJ13679 were the same as that of AJ1511.

Table I. Bacterial Strains and Plasmids Used in This Study

Strains and Plasmids	Relevant Characteristics
Corynebacterium gluatmicum AJ1511 (ATCC13869)	wild type strain
Corynebacterium glutamicum AJ13678	transformant of AJ1511 with pGDH
Corynebacterium glutamicum AJ13679	transformant of AJ1511 with pICDH
pGDH	*gdh* (10), Cm^r, ori
pICDH	*icd* (11), Km^r, ori

Medium with following composition was used for the preculture and main culture (per liter deionized water): 80 g glucose, 30 g $(NH_4)_2SO_4$, 10 g KH_2PO_4,

42

0.4 g $MgSO_4 \cdot 7H_2O$, 0.01 g Fe $SO_4 \cdot 7H_2O$, 0.01 g Mn $SO_4 \cdot 5H_2O$, 200 µg vitamin B1 · HCl, 3 µg biotin, 480 mg-N soybean protein hydrolysate.

Cultivation Conditions
All cultivations were done in batch mode. A 5L jar-fermentor (KMJ-5B, Mitsuwa, Japan) with a liquid working volume of 3L was used for fermentation. The temperature was controlled at 31.5°C and pH was maintained at 7.20 by addition of 28% ammonia water. Dissolved oxygen concentration was maintained at higher levels than 3.0 mg/L by manipulating agitation speed in the range of 100 – 700 revolution per minute. Aeration rate was kept at one vvm.

Online Measurement
The data of CO_2 and O_2 concentrations in exhaust gas were measured by gas analyzers (CO_2 analyzer, Fuji Electric Co., Japan and O_2 analyzer, Fuji Electric Co., Japan), respectively. Cell concentration was measured by laser turbidimeter (LA-300LT, ASR Co., Japan). Addition rate of ammonia water was monitored by electric balance (FX-6000, Kensei Kogyo Co. Inc., Japan). All these data were input to a data logger (Thermic, Eto Electric Co., Japan) and transferred to the personal computer (PC-286VE, Epson Co., Japan) by RS-232C multiplexer.

Offline Measurement
Cell concentration was measured by OD_{660}. Glucose and glutamate concentrations were measured enzymatically by glucose analyzer (Model-2700, YSI, USA) and Bioanalyzer (BF-400, Oji Electric Co., Japan). To determine enzyme activities, cells were washed twice with 0.2 N KCl and resuspended in Tris-HCl buffer. Cells were disrupted by sonication and the supernatant was used as cell extract. The enzymes activities were assayed photometrically at 31.5°C in 1 mL of a reaction mixture. The reaction mixture for ICDH activity assay contained 35mM Tris-HCl, 1.5mM $MnSO_4$, 0.1mM $NADP^+$, 1.3mM isocitrare. The ICDH activity was measured from increase in the absorption at 340 nm. (11).

The reaction mixture for GDH activity assay contained 100mM Tris-HCl, 20mM NH_4Cl, 0.25mM NADPH, 10 mM 2-oxoglutarate. The GDH activity was measured from decrease in the absorption at 340 nm. (10).

The reaction mixture for ODHC activity assay contained 100mM N-tris(hydroxymethyl)methyl-2-aminoethanesulfornic acid (TES)-NaOH, 0.2 mM Co-enzyme A, 0.3m M thiamine pyrophosphate, 1mM 2-oxoglutarate, 3mM cysteine, 5mM $MgCl_2$, and 1mM 3-acetylpyridine adenine dinucreotide, the last of which was used instead of NAD^+. The enzyme activity was measured from increase in the absorption at 365 nm (8). The protein concentration was measured by protein assay kit (Biorad Co., USA).

Metabolic Reaction Model

Metabolic map for glutamate synthesis is shown in Figure 1. Stoichiometric representations are summarized in TableII.

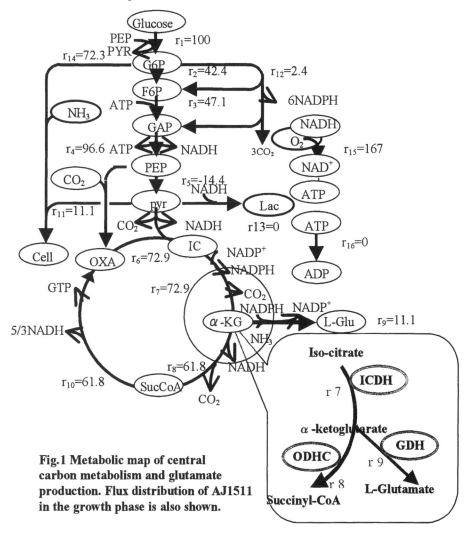

Fig.1 Metabolic map of central carbon metabolism and glutamate production. Flux distribution of AJ1511 in the growth phase is also shown.

Table II. Metabolic Reaction Model for Glutamate Production

(Phosphotransferase system)
r_1: Gluc +PEP=>G6P+pyr
(Glycolysis)
r_2: G6P=>F6P
r_3: F6P+ATP=>2GAP
r_4:GAP=>PEP+ATP+NADH
r_5: PEP=>pyr +ATP
(TCA cycle and glutamate synthesis)
r_6: pyr+Oxa=>IC+CO_2+NADH
r_7: IC=>KG+CO_2+NADPH
r_8: KG=>SucCoA+CO_2+NADH
r_9: KG+NH_3+NADPH=>Glu
r_{10}: SucCoA=>Oxa+5/3NADH+ATP
(Anaplerotic pathway)
r_{11}: PEP+CO_2=>Oxa
(Pentose phosphate pathway)
r_{12}: 3G6P=>2F6P+GAP+3CO_2+6NADPH
(Lactate synthesis)
r_{13}: pyr+NADH=>Lac
(Anabolism)
r_{14}: $\beta\alpha$ G6P+β(1-α) pyr+NH_3+(DW/Y_{ATP})+1.06NADPH=> $C_{4.17}H_{7.32}O_{1.92}N$
(Oxidative phosphorylation)
r_{15}: NADH+O_2=>2(P/O)ATP
(Excess ATP)
r_{16}: ATP=>ADP

Abbreviation: Gluc, glucose; G6P, glucose-6phosphate; PEP, phosphenolpyruvate; pyr, pyruvate; F6P, fructose-6phosphate; GAP, glycelardehyde-3-phosphate; IC, isocitrate; KG, 2-oxoglutarate; SucCoA, succinyl CoA; Oxa, oxaloacetate; Lac, lactate

In this model, it was assumed that GTP and FADH were equivalent to ATP and 2/3NADH, respectively. Parameters α and β were to be 0.8 and 0.76, respectively. (P/O), Y_{ATP}, and molecular weight of the cell, DW, were to be, 2.0, 10.0 (g-cell/mol-ATP), and 107.5 g/moll-cell, respectively. Glyoxylate shunt was ignored because the activity of isocitrate lyase at the entrance of glyoxlate shunt was very low under glucose rich condition. Glutamine synthetase-glutamine-oxoglutarate aminotransferase (GS-GOGAT) system was also ignored because the GDH was main pathway to NH_3 uptake under nitrogen rich condition in the Corineform bacteria.

The material balances of intracellular and extracellular metabolites are represented in eq.(1).

$$Ar_c = \bar{r}_m \qquad (1)$$

where r_c (16x1), and $\bar{r}_m$ (19x1) are the calculated reaction rate vector and measured reaction rates with measured errors vector represented as

$$r_c = [r_1 \quad r_2 \quad r_3 \quad \cdots \quad r_{16}]^T \qquad (2)$$

and

$$\bar{r}_m = [r_{gluc} \ r_{cell} \ r_{Glu} \ r_{CO2} \ r_{O2} \ r_{NH3} \ r_{Lac}$$
$$r_{G6P} \ r_{F6P} \ r_{GAP} \ r_{PEP} \ r_{py} \ r_{IC} \ r_{KG} \ r_{SuccoA} \ r_{Oxa} \ r_{ATP} \ r_{NADH} \ r_{NADPH}]^T \qquad (3)$$

respectively. A is a stoichiometric coefficients matrix (19x16).

Metabolic Flux Analysis (MFA)

Reaction rates of glucose uptake, cell growth, glutamate synthesis, CO_2 evolution, O_2 uptake, ammonia uptake, lactate synthesis were measured, and intracellular metabolites accumulation rates were to be zero because of pseudo steady state assumption. Since the system has redundancy of measurement (degree of freedom is three), r_c is determined with reconciled measured values by a least square method (9) as

$$r_c = [A^T A]^{-1} A^T \hat{r}_m \qquad (4)$$

where $\hat{r}_m$ is a reconciled measured reaction rate vector given by $\bar{r}_m$ and the estimated measured error vector, $\hat{\delta}$ as

$$\hat{r}_m = \bar{r}_m - \hat{\delta} \qquad (5)$$

where $\hat{\delta}$ is determined with the information of variance-covariance matrix of measured errors. The consistency of the metabolic reaction model was checked by comparison the consistency index of the developed model, h, with χ^2 statistics value of the degree of redundancy of the system (9).

Determination of Flux Control Coefficients (FCCs) Based on the Large Perturbation Theory

Flux control coefficients, FCCs, C_i^k, are defined as relative effects of infinitesimal changes in the activities of enzymes on changes in the steady state fluxes.

$$C_i^k = \frac{\partial J_k / J_k}{\partial e_i / e_i} \tag{6}$$

where e_i and J_k are enzyme activity of reaction i and flux of reaction k, respectively. At the branch point such as 2-oxoglutarate in glutamate production, nine FCCs are defined because there are three enzyme activities and three fluxes at the branch point. To determine FCCs experimentally, many methodologies have been proposed (9). Among them a bottom up approach was based on the kinetic model of the metabolic pathway, which is very robust if the kinetic model represents the real metabolic pathway. However, there exist difficulties to construct the good kinetic model because it is necessary to identify many kinetic parameters in the model. An approach to draw the figure of relationship between e_i and J_k experimentally, also encounters some difficulties because many different plots of e_i and J_k are necessary and this fact requires a lot of genetically engineered strains. A large perturbation theory is a method based on the linear reversible enzyme reactions network, and this method is available for analysis of branch point, using small number of perturbation experiments of enzyme activities and metabolic flux distribution data (5, 6, 12).

Determination Method of FCCs with Fluxes and Enzyme Activities Data

In the large perturbation experiments at the branch point, first deviation index, D_i^k, are determined experimentally as

$$D_i^k = \frac{\Delta J_k / J_k^r}{\Delta e_i / e_i^r} \tag{7}$$

where ΔJ_k, J_k^r, Δe_i, and e_i^r are large perturbation of flux of k reaction, perturbed flux of k reaction, large perturbation of enzyme activity of i reaction, and perturbed enzyme activity of i reaction, respectively.

FCCs are determined by deviation index with the correction factors F_i^k as

$$C_i^k = D_i^k F_i^k \tag{8}$$

F_i^k are given as (9, 12)

$$F_i^k = \frac{1 - C_i^i \dfrac{r_i - 1}{r_i}}{1 - D_i^k \dfrac{r_i - 1}{r_i}} \tag{9}$$

where r_i are enzyme activity amplification factors defined as the ratio of the perturbed enzyme activity to the original enzyme activity of the reaction i as

$$r_i = \frac{e_i^r}{e_i^0} \tag{10}$$

As for the reaction directly catalyzed by the perturbed enzyme FCC is exactly coincided with the deviation index as

$$C_i^i = D_i^i \tag{11}$$

or

$$F_i^i = 1 \tag{12}$$

After FCCs are determined by above equations, flux change is predicted by the specified change in enzyme activity. Flux amplification factor is defined as the ratio of perturbed flux to the original flux, f_i^k, as

$$f_i^k = \frac{J_k^r}{J_k^0} \tag{13}$$

The flux amplification factors is calculated from the information of determined FCCs and enzyme amplification factors as

$$f_i^k = \frac{1 - \left(C_i^i - C_i^k\right)\frac{r_i - 1}{r_i}}{1 - C_i^i \frac{r_i - 1}{r_i}} \tag{14}$$

From the determined flux amplification factors, it is predicted how flux will be changed when the new genetic modification is introduced in the metabolic pathway.

Results and Discussion

Metabolic Flux Analysis by The Constructed MR Model

The MR model shown in Table II and eq. (1) was constructed, and consistency of the model was checked by using the consistency index. When the ammonia addition rate for pH control was assumed to be ammonia uptake rate, large residual error was observed and the MR model was not consistent. Taking into account decrease in pH due to accumulation of glutamate in the medium, the estimated values of ammonia uptake rate were corrected. The model with the corrected measured values of ammonia uptake rate was consistently satisfied by the χ^2 statistics. Here after this model was used for flux distribution analysis. As an example, flux distribution of wild type strain AJ1511 in the growth phase was shown in Fig.1.

Metabolic Flux Distribution at the 2-Oxoglutarate Branch Point in Perturbed Experiments

Metabolic flux distribution was analyzed in all the strain of AJ1511, AJ13678 and AJ13679. In this article the flux distribution at the key branch point of 2-oxoglutarate is focused on. By using the strains of AJ13678 and AJ13679 and biotin depletion condition, respectively, effects of the amplification and attenuation of enzyme activities on the glutamate flux were analyzed. The flux distribution at the 2-oxoglutarate is shown in Table III and the amplification factors based on the flux distribution analysis was summarized in Table IV.

Table III. Comparison of Normalized Fluxes at 2-Oxoglutarate

	WT^a	*ICDH enhanced*[b]	*ODHC attenuated*[c]	*GDH enhanced*[d]
J_7(ICDH)	0.73	0.93	0.68	0.83
J_8(ODHC)	0.62	0.66	0.15	0.71
J_9(GDH)	0.11	0.27	0.53	0.11

a: fluxes of wild type strain AJ1511 were estimated before biotin depletion.
b: fluxes of *icd* enhanced strain of AJ13679 were estimated.
c: fluxes after ODHC attenuation of wild type strain were estimated after biotin depletion with both experimental data of AJ1511 and AJ13679.
d: fluxes of *gdh* enhanced strain of AJ13678 were estimated.

Table IV. Enzyme Activity Amplification Factors r_i and Flux Amplification Factors f_i^k in ICDH, ODHC, and GDH perturbed experiments

	ICDH enhanced[a]	*ODHC attenuated*[b]	*GDH enhanced*[c]
r_i	2.96	0.52	3.21
f_i^k			
J_7(ICDH)	1.27	0.93	1.13
J_8(ODHC)	1.06	0.24	1.15
J_9(GDH)	2.41	4.77	1.02

a: AJ13679 (ICDH enhanced) compared with those of AJ1511 (wild type).
b: before biotin depletion were compared with after biotin depletion of AJ1511
c: AJ13678 (GDH enhanced) compared with those of AJ1511.

It was found that enhancement of ICDH and GDH did not change flux distribution at the branch point of 2-oxoglutarate. Even though ICDH and GDH activities were enhanced 2.96 and 3.21, respectively, more than 70 % of carbon flux still flows into TCA cycle. On the other hand, when ODHC activity was decreased to around 50 % after depletion of biotin, dramatic changes in fluxes of GDH and ODHC were given. More than 75 % carbon directed into glutamate production. Trend of time courses of enzyme activities, glutamate production and growth was almost the same as those reported previously (8).

By using the data of Table IV, Deviation index and FCCs were determined, and calculation results are summarized in Table V.

Table V. Deviation Index D_i^k and FCCs C_i^k at 2-Oxoglutarate

	ICDH enhanced	ODHC attenuated	GDH enhanced
D_i^k			
J_7(ICDH)	0.32	0.08	0.17
J_8(ODHC)	0.09	3.43	0.19
J_9(GDH)	0.88	-0.84	0.025
C_i^k			
J_7(ICDH)	0.32	0.33	0.19
J_8(ODHC)	0.08	3.43	0.22
J_9(GDH)	1.67	-16.9	0.025

The FCCs of ODHC attenuation on glutamate production became negative value because glutamate production was enhanced by decrease in ODHC activity, namely competitive pathway showed negative sensitivity. The most important issue in Table V is that the greatest impact on glutamate production indicates the greatest absolute value of the FCC. The greatest value of FCC is one of the ODHC attenuation. Second impact on glutamate production was given by ICDH enhancement, and GDH enhancement was not effective for glutamate production. These flux distribution analysis and FCCs data quantitatively supported that GDH and ICDH enhancements did not change glutamate production effectively (10, 11) but ODHC attenuation after biotin depletion was a trigger of glutamate production (8).

As shown in Table III and IV, when GDH and ICDH were enhanced glutamate production was not changed significantly. Especially, GDH did not make any impact on glutamate flux. This is a clear evidence that the GDH does not have large responsibility for glutamate flux. From the view point of kinetics, it is speculated that Km value of GDH for 2-oxoglutarate concentration is very large and the elasticity of this enzyme is very large. FCCs values in Table V indicated the grades of impact of changes in enzyme activities on fluxes.

Summation of all FCCs for the enzyme activity changes on each flux was not unity and this result would become a controversial problem. One reason is clearly that there are no constraints related with summation theorem in the FCCs determination method used here. This fact suggested that there are other enzymes which did not exist around 2-oxoglutarate had possibility to control metabolic network flux. Upstream of ICDH would have the responsibility to determine the glutamate flux. Alternative method to determine FCCs with summation theorem constraints has been also developed (6). The results by this method strongly suggested upstream of ICDH has more responsibility. This results will be published in the near future.

Prediction of Flux Change from Change in Enzyme Activity
Flux change from change in the specified enzyme activity was estimated by the determined FCCs shown in Table V. The prediction results are summarized in Table VI. The flux amplification factors were predicted as example cases when ICDH, ODHC, and GDH activities were changed 5 times, 1/5 times, and 5 times, respectively.

Table VI. Predicted Flux Amplification Factors f_i^k at 2-Oxoglutarate

	ICDH 5 x enhanced	ODHC 1/5 x attenuated	GDH 5 x enhanced
f_i^k			
J_7(ICDH)	1.35	0.91	1.16
J_8(ODHC)	1.08	0.06	1.18
J_9(GDH)	2.80	5.60	1.02

The 5.6 times larger glutamate flux is given by the 1/5 attenuated ODHC activity, comparing with the original strain. This modification should be selected as the best candidate to improve glutamate production in *Corynebacterium glutamicum*.

References

1. Bailey, J.E. Towards a science of metabolic engineering. *Science* **1991**, *252*, 1668-1675.
2. Stephanopoulos, G., Vallino J.J. Network rigidity and metabolic engineering in metabolite over production. *Science* **1991**, *252*, 1675-1681.
3. Vallino J.J., Stephanopoulos, G. Metabolic flux distributions in *Corynebacterium glutamicum* during growth and lysine over production. *Biotechnology Bioengineering* **1993**, *41*, 633-646.
4. Marx, A., de Graaf, A.A., Wiechert, W., Eggeling, L., Sahm, H. Determination of the fluxes in the central carbon metabolism of *Corynebacterium glutamicum* by nuclear magnetic resonance spectroscopy combined with metabolite balancing. *Biotechnology Bioengineering* **1996**, *49*, 111-129
5. Small, J.R., Kacser, H. Responce of metabolic systems to large changes in enzyme activities and effectors 1. The linear treatment of unbranched chains. *European Journal of Biochemistry*, **1993**, *213*, 613-624.
6. Stephanopoulos, G., Simpson, T., Flux amplification in complex metabolic network. *Chemical Engineering Science*, **1997**, *52*, 2607-2627.
7. Takiguchi, N., Shimizu, H., Shioya, S. An on-line physiological state recognition system for the lysine fermentation process based on a metabolic reaction model. *Biotechnology Bioengineering*, **1997**, *55*, 170-181.
8. Kawahara, Y., Takahashi-Fuke, K., Shimizu, E., Nakamatsu, E., Nakamori, S. Relationship between the glutamate production and the activity of 2-oxoglutarate dehydrogenase in *Brevibacterium lactofermentum*. *Bioscience, Biotechnology, and Biochemystry*, **1997**, *61*, 1109-1112.
9. Stephanopoulos, G., Aristidou, A., Nielsen, J. Metabolic Engineering: Academic Press, San Diego, CA, 1998.
10. Eikmanns, B.J., Rittmann, D., Sahm, H. Cloning, sequence analysis, expression, and inactivation of the *Corynebacterium glutamicum icd* gene encoding isocitrate dehydrogenase and biochemical characterization of the enzyme. *J. Bacteriology*, **1995**, *177*, 774 -782.
11. Bormann, E.R., Eikmanns, E.J., Sahm, H. Molecular analysis of the *Corynebacterium glutamicum gdh* gene encoding glutamate dehydrogenase, *Molecular Microbiology*, **1992**, *6*, 317-326.
12. Small, J.R., Kacser, H. Responce of metabolic systems to large changes in enzyme activities and effectors 1. The linear treatment of branched pathways and metabolite concentrations. Assessment of general no-linear case. *European Journal of Biochemistry*, **1993**, *213*, 625-640.

Chapter 5

Biological Production of Xylitol by *Candida tropicalis* and Recombinant *Saccharomyces cerevisiae* Containing Xylose Reductase Gene

Kwan-Hoon Moon[1], Woo-Jong Lee[1], Jay-Han Kim[1], Jin-Ho Choi[1], Yeon-Woo Ryu[2], and Jin-Ho Seo[1,*]

[1]Department of Food Science and Technology, School of Agricultural Biotechnology, Seoul National University, Suwon 441–744, Korea
[2]Department of Molecular Science and Technology, Ajou University, Suwon 442–749, Korea

Xylitol, a natural sweetener, was produced from xylose using *Candida tropicalis* ATCC 13803 and recombinant *Saccharomyces cerevisiae* containing the xylose reductase gene from *Pichia stipitis* in various culture modes. A two-substrate fermentation was designed in order to increase xylitol yield and volumetric productivity for *C. tropicalis*: glucose was used for cell growth and xylose for xylitol production. Computer simulation was undertaken to optimize the two-substrate fermentation using kinetic equations describing rates of cell growth and xylose bioconversion as a function of ethanol concentration. The optimized two-substrate fermentation resulted in xylitol yield of 0.81 g-xylitol/g-xylose and volumetric productivity of 5.06 g-xylitol/L·hr, which are in good agreement with the computor simulation results. To improve xylitol productivity and final xylitol concentration without

© 2002 American Chemical Society

53

sacrificing xylitol yield, cell-recycle fermentations were attempted. A series of cell-recycle experiments showed that the feeding of xylose, glucose and yeast extract in the xylitol production phase was the most effective in enhancing xylitol productivity. A metabolically engineered *Saccharomyces cerevisiae* containing a xylose reductase gene from *Pichia Stipitis* was employed in an attempted to improve xylitol yield further. The recombinant *S. cerevisiae* strain in the optimized fed-batch culture resulted in xylitol yield of 0.95 g-xylitol/g-xylose and xylitol productivity of 1.69 g-xylitol/L·hr.

Introduction

Xylitol, a five-carbon sugar alcohol, is a reduction product of xylose and found naturally in plants, fungi, microorganisms, and the human body (in which the liver alone can synthesize 5 to 15 g of xylitol daily) (*1*). It has a sweetness equivalent to sucrose with an extreme cooling effect and has been used as a sweetening agent in human foods (*2, 3*). Recently, it has been gaining significance as an alternative sweetener due to its functional properties such as reducing the development of dental caries (cavities) and insulin non-dependency in metabolic regulation. It can be used as a sugar substitute by diabetics and glucose-6-phosphate dehydrogenase deficient individuals since it does not require insulin and glucose-6-phoaphate dehydrogenase for regulation of metabolism (*4, 5*). Xylitol is currently produced by chemical hydrogenation of xylose in hemicellulose hydrolyzates using Ni/Al_2O_3 as a catalyst. The product cost is high due to difficulties of purification and separation of xylitol, removal of by-products from hemicellulose hydrolyzates and a low yield of 40~50 % based on xylan (*6*). Even though enzymatic conversion of xylose to xylitol is possible by using NAD(P)H-dependent xylose reductase from a technical aspect, the enzymatic process is not feasible from an economic standpoint as the continuous regeneration of NAD(P)H is required. Instead, microbial conversion for xylitol production has been employed by using *Candida* species such as *C. pelliculosa* (*7*), *C. boinidii* (*8*), *C. guilliermondii* (*9*), *C. parapsilosis* (*10, 11*) and *C. tropicalis* (*12, 13*). The microbial process has several advantages compared with the chemical process in terms of selective conversion of xylose to xylitol with high yield. In the xylose metabolism of *C. tropicalis*, xylose was taken up by a specific transferase and reduced to xylitol by xylose reductase (XR) with NADPH followed by conversion to xylulose by xylitol dehydrogenase (XDH) with NAD^+ (Figure 1). Xylulose is then used for cell growth and

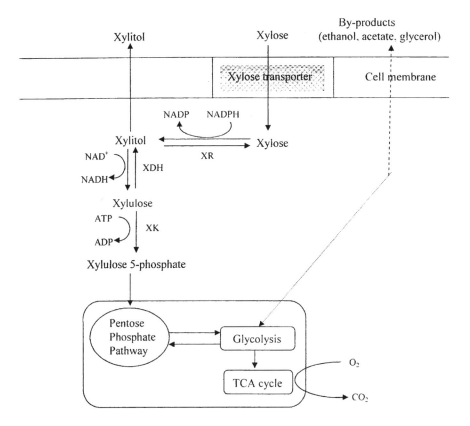

Figure 1. Metabolic pathway of xylose in C. tropicalis

NADPH regeneration through the pentose phosphate pathway after conversion to xylulose-5-phosphate by xylulose kinase with ATP as a cofactor. To obtain high xylitol yield, the xylose flux to xylulose has to be controlled by oxygen supply sufficient for regeneration of NADPH and cell maintenance. Low oxygen levels are also required for xylitol production because they decrease the NAD^+/NADH ratio, which favors the xylitol dehydrogenase-catalyzed reaction to xylitol accumulation by changing the equilibrium constant.

For the continuous production of xylitol from xylose, immobilization techniques of whole cells have been used (*14*). Calcium alginate, polyacrylamide, and non-woven fabric were used as the entrapping materials of cell immobilization. Cell immobilization techniques could improve the productivity, but were associated with a lower mass transfer rate from a bulk flow to cells, and leakage of cells into the medium (*15*). In order to solve the problems involved in the cell immobilization system, a cell-recycle fermentation system with a hollow fiber membrane was used to keep high concentration of cells as biocatalysts for conversion of xylose to xylitol. Specifically, the composition of the feeding solution in the xylitol production period was optimized for high productivity.

Natural xylose-fermenting yeasts utilize a fraction of xylose for cell growth and endogenous metabolism, resulting in decreased xylitol yield. A recombinant *S. cerevisiae* strain transformed with the xylose reductase gene (*XYL1*) could allow the efficient conversion of xylose to xylitol with high yields over 95 % as xylitol is not further metabolized to xylulose (*16*). Wild type *S. cerevisiae* is known to lack the metabolic pathway to convert xylose to xylulose via xylitol. A metabolically engineered *S. cerevisiae* strain containing the xylose reducing gene from *P. stipitis* was analyzed in order to produce xylitol from xylose. A two-substrate fermentation process has been adopted: glucose is used as an energy source for cell growth and co-factor regeneration and xylose as a substrate for conversion to xylitol.

This paper summarizes research efforts on xylitol production which have been performed in my laboratory, to compare the fermentation characteristics of wild type *C. tropicalis* and recombinant *S. cerevisiae*.

Materials and Methods

Microorganism and Culture Conditions

C. tropicalis ATCC13803 was maintained at 4 °C on a YPX agar plate containing 10 g/L yeast extract, 20 g/L bactopepton, 20 g/L xylose and 15 g/L agar. The medium composition for inoculation and main culture was the same as the maintenance medium except for carbohydrate concentrations. Twenty g/L glucose and 60 g/L xylose were added to the preculture medium and 100 g/L xylose and various concentrations of glucose were added to the fermentation medium. This yeast was cultured in 100 mL of the preculture medium at 30 °C, pH 6 and 200 rpm in a shaking incubator (Vision, Seoul, Korea). Batch and cell-recycling fermentations were performed at 30 °C, pH 6, 500 rpm and 1.0 vvm (K_{La} = 1.06 min^{-1}) in a 2.5-L jar fermentor (Korbiotech, Incheon, Korea) containing 1 L of the fermentation medium. Hollow fiber membrane (Amicon cartridge type HIP 100-43, USA) with molecular weight cut off 100,000 was installed for cell recycle fermentations.

A recombinant yeast strain able to express the xylose reductase gene from *P. stipitis* was constructed using *S. cerevisiae* EH13.15 [*Mat* α, *trp*1] as a host harboring plasmid pY2XR (Figure 2). The xylose reductase gene in plasmid pY2XR is expressed by the constitutive glyceraldehyde dehydrogenase promoter. Inoculum was prepared by transferring a loopful of recombinant *S. cerevisiae* cells grown on the YNB plate to a test tube containing 5 mL of YNB medium. Flask cultures were carried out in 500-mL baffled flasks containing 100 mL culture media. Three mL of the inoculum culture was transferred and incubated at 30 °C in a shaking incubator at 200 rpm. Fermentor cultures were carried out in a 2.5-L fermenter with a 1-L working volume. An agitation speed was set at 400 rpm, and aeration rate at 1 vvm. pH was controlled at 5.0. Fed-batch fermentations were done with 750 mL of the defined medium as a start-up medium described by O'Connor *et. al.*(*17*). After glucose in the medium was depleted, a mixture of glucose and xylose was fed to study the effect of various molar ratios of xylose to glucose during the fed-batch period.

Analytical Methods

Xylose, xylitol and glucose were determined by HPLC (Knauer, Berlin, Germany) using the Carbohydrate Analysis (Waters, Milford, U.S.A.) column with 85 % (v/v) acetonitrile as mobile phase at a flow rate of 2 mL/min. Carbohydrates were measured by using a reflective index detector (Knauer). Aminex HPX-87C and Aminex HPX 87H (BioRad, Hercules, CA, U.S.A.) columns were also used to monitor by-products. Xylitol concentrations below 1 g/L were determined by using the D-sorbitol/xylitol kit (Boehringer Mannheim, Mannheim, Germany). Ethanol was measured by gas chromatography (Younglin, Seoul, Korea) using a column (Carbowax 20M, Hewlett Packard, U.S.A) with N_2 as a carrier gas at a flow rate of 50 mL/min and a flame ionized detector. Temperatures of injector, detector, and column were 200 °C, 200 °C, and 150 °C, respectively. Cell mass was estimated by using the relationship between dry cell weight and optical density (OD) measured at 600 nm. Specific rates of xylose consumption and xylitol production were defined as differences in xylose and

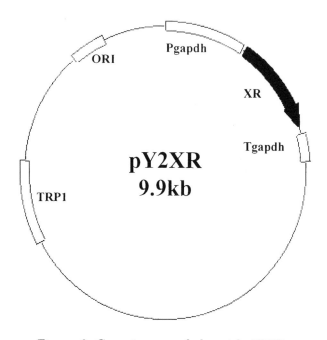

Figure 2. Genetic map of plasmid pY2XR

(Reproduced with permission from reference 22. Copyright 2000 ELSEVIER)

xylitol concentrations divided by average cell mass and the time interval between the two samples of interest, respectively.

Results and Discussion

Two-Substrate Batch Culture of *C. tropicalis*

Efficient production of xylitol from xylose requires continuous regeneration of NADPH, a cofactor of xylose reductase. Under aerobic conditions NADPH is normally produced by glucose-6-phosphate dehydrogenase and 6-phosphogluconate dehydrogenase in the pentose phosphate pathway, by isocitrate dehydrogenase in the TCA cycle and transhydrogenase, the exchanger of H^+ from NADH to $NADP^+$ in the cytosol (*18, 19*). All these pathways are related to energy generation for cell growth and redox balance. The best way to increase xylitol yield would be simultaneous transport of a co-substrate with xylose into the cell. The co-substrate is used primarily for NADPH regeneration while xylose is converted to xylitol without being metabolized further. To make such a scheme possible, xylose and co-substrate must enter the cell at the same time without inhibition of the required transferases and do not inhibit enzymes involved in co-substrate metabolism and xylose conversion. Alcohols, hexoses or pentoses are not suitable as co-substrate owing to the inhibition of cell metabolism and high costs. Glucose, a good candidate for a co-substrate, blocks xylose transport and represses XR activity. Therefore a certain amount of xylose must be used for NADPH regeneration, which decreases xylitol yield. To increase xylitol yield, the xylose flux to cell mass has to be minimized, but provides sufficient maintenance energy and NADPH regeneration by controlling the oxygen supply rate. Microaerobic conditions might keep the NAD^+/NADH ratio and ATP levels low. The xylose flux to cell growth is restricted by the lack of cofactors necessary for xylitol dehydrogenase and xylulose kinase. Glucose was chosen as co-substrate for cell growth to obtain high volumetric productivity. Since volumetric productivity is proportional to cell mass, it is necessary to increase cell mass by using glucose as an energy source. A two-substrate fermentation is established in such a way that glucose is used for cell growth and xylose is converted to xylitol with high yield.

A number of two-substrate fermentations using *C. tropicalis* were performed to see the pattern of production and utilization of by-products (*20*). Major by-products of xylose and glucose metabolism included ethanol, acetic acid and glycerol which were usually utilized again as substrates for cell growth. Only ethanol was produced from glucose without formation of glycerol and acetic acid and was not consumed during the xylose bioconversion phase. Since the xylose flux to glycolysis and oxygen were limited, no ethanol was produced from xylose by the Crabtree effects during the xylose bioconversion phase, which is beneficial to high xylitol yield. After complete depletion of glucose, xylose was converted to xylitol with 81 % yield. High glucose concentrations increased volumetric productivity by reducing conversion time due to high cell mass, but also led to production of ethanol, which, in turn, inhibited cell growth and xylitol production. Computer simulation was undertaken to optimize an initial glucose concentration using kinetic equations describing rates of cell growth and xylose bioconversion as a function of ethanol concentration. Kinetic constants involved on the equations were estimated from the experimental results. The optimum initial glucose concentration to maximize volumetric productivity was estimated to be 32 g/L, a total fermentation time of 16.8 hr and volumetric productivity of 5.15 g-xylitol/L·hr. Simulation results were experimentally verified in the two-substrate fermentation under the same conditions. As shown in Figure 3, the xylose bioconversion phase was clearly separated from the cell growth phase. After depletion of glucose, xylose was converted to xylitol and ethanol produced from glucose was not utilized during the xylose consuming period. A xylitol yield of 0.81 g-xylitol/g-xylose and a volumetric productivity of 5.06 g-xylitol/L·hr were obtained, which was in good agreement with the predicted values of computer simulation.

Cell-Recycle Fermentation using *C. tropicalis*

To improve xylitol productivity and final xylitol concentration without sacrificing xylitol yield, cell recycle fermentations were attempted (*21*). The fermentation medium was filtered through hollow fiber membrane and the cells were recycled back to the fermentor to maintain high cell concentrations for conversion of xylose to xylitol. Fresh medium was fed to make up the fermentation broth. A number of cell-recycle fermentations were performed to determine the best composition of feeding solution during the cell-recycle period.

First, a cell-recycle fermentation with feeding of xylose only was carried out. Xylose solution (750 g/L) was added into the culture broth when the xylose concentration in the medium fell below 1 %. The concentration of xylose was adjusted from 1 % to 10 % after transition to the cell recycle fermentation mode. Both xylitol productivity and yield decreased to 2.46 g-xylitol/L·hr and 0.75 g-xylitol/g-xylose, respectively compared with the batch fermentation even with higher concentration of xylitol, 207 g/L. Interestingly, xylulose, an isomer of xylose and an intermediate in xylose metabolism, was detected at the end of the fermentation, suggesting shortage of nutrients and intracellular energy.

Another cell-recycle fermentation was performed by feeding xylose (750 g/L) and yeast extract (100 g/L) as a nitrogen source to supply more nutrients. As expected, xylulose was not detected, and both yield and productivity of xylitol increased slightly compared with a former cell-recycle fermentation. Xylitol at 244 g/L was obtained, corresponding to a 2.8-fold enhancement

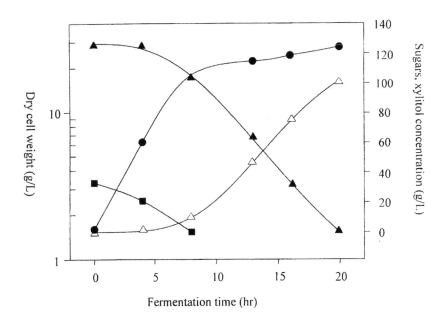

*Figure 3. Experimental results of the two-substrate batch
culture of C. tropicalis under the optimized
conditions (● Dry cell weight (g/L), ■ glucose
(g/L), ▲ xylose (g/L), △ xylitol (g/L))*
(Reproduced with permission from reference 20. Copyright 1999 Nature
Publishing Group.)

compared with the batch fermentation. However, xylitol productivity did not improve much compared with the batch fermentation.

Finally, the cell-recycle fermentation with feeding of xylose, yeast extract, and glucose as a co-substrate was performed (Figure 4). Glucose solution (200 g/L) was added continuously into the fermentor at 8 mL/hr after the initial glucose added was depleted. Glucose was not detected even in the cell-recycle period. This operation strategy was designed to supply a carbon source for continuous regeneration of NAD(P)H. The optimized cell-recycle fermentation resulted in xylitol productivity of 4.94 g-xylitol/L·hr and final xylitol concentration of 189 g/L with the almost same xylitol yield as before. These experimental results are equivalent to a 2.2-fold enhancement in xylitol concentration and a 1.3-fold increase in volumetric xylitol productivity compared with the two-substrate batch fermentation. Also, xylitol yield slightly increased to 0.82 g-xylitol/g-xylose. It was concluded from these results that the cell-recycle fermentation with feeding of xylose, yeast extract, and glucose was effective for xylitol production by *C. tropicalis.*

Fed-Batch Fermentation of Recombinant *Saccharomyces cerevisiae* **containing xylose reductase gene**

In the previous experimentals, *C. tropicalis* utilized a xylose for cell growth and endogeneous metabolism, resulting in xylitol yield of 0.81 g-xylitol/g-xylose. As xylose is an expensive raw material, a metabolic engineering approach was attempted in order to improve xylitol yield. *S. cerevisiae* was chosen as a host since the yeast is known to lack the metabolic activity for conversion of xylose to xylitol. A recombinant *S. cerevisiae* was constructed by introducing the xylose reductase gene (XR) form *P. stipitis.* The recombinant yeast in a batch culture was able to produce xylitol from xylose with over 95 % yield. Fed-batch fermentations are performed to find out the optimum fermentation strength for maximum xylitol productivity and final xylitol concentration (*22*). At the stationary phase of batch cultures, xylitol production ceased upon depletion of glucose, which is an essential co-substrate for supplying energy sources necessary for endogenous metabolism and co-factor regeneration. A two-substrate fed-batch fermentation process was adopted for simultaneous addition of substrate (xylose) and co-substrate (glucose) to avoid co-substrate depletion during the bioconversion period. A key factor in optimizing fed-batch fermentations was to maintain a high ratio of xylose to glucose in the fermentation medium. Glucose and xylose are also known to be transported by the membrane transporter through facilitated diffusion. They appeared to share the same membrane transporter and competitively inhibited the transport of each other (*23, 24*). However, as the membrane transporter has much higher affinity for glucose than xylose, glucose is more efficiently transported into the cell.

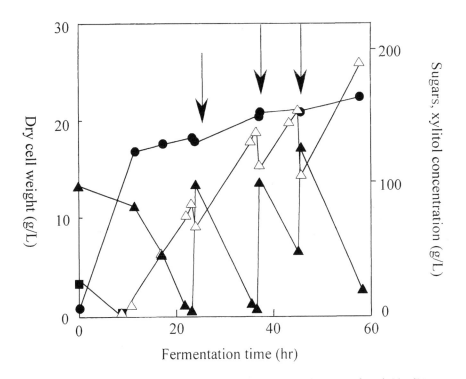

Figure 4. Cell-recycle fermentation of C. tropicalis with feeding xylise (750 g/L), yeast extract (100 g/L), and glucose (200 g/L) at 8 mL/hr. (*The arrows indicate the feeding time of make-up solution.*)

(● *Dry cell weight (g/L)*, ■ *glucose (g/L)*, ▲ *xylose (g/L)*, Δ *xylitol (g/L)*) (Reproduced with permission from reference 21. Copyright 2000 Kluwer Academic Publisher)

Such a difference in affinity may result in ethanol accumulation and low xylitol productivity in the two-substrate fed-batch fermentation. Therefore, the feed rates of glucose and xylose had to be controlled in the fed-batch operation phase to keep the ratio of xylose to glucose as high as 40. Glucose concentration was maintained around 0.35 g/L throughout the fermentation; hence, ethanol productivity decreased to 0.36 g/L·hr (Figure 5). Cell mass increased to 12.7 g/L with xylitol productivity of 1.69 g/L·hr. It seemed that a sufficient amount of glucose, required for cell growth and co-factor regeneration, was transported even under such a severe glucose-limited condition. Glucose feeding was stopped when ethanol concentration reached 20.8 g/L and the final xylitol concentration was 105.2 g/L.

Conclusions

Xylitol is a value-added material produced from xylose by hydrogenation. Various biological conversion processes applied to produce xylitol. The first investigation was undertaken to produce xylitol using a wild type *C. tropicalis* in two-substrate batch fermentation. Glucose was used to supply energy source at cofactors and xylose was utilized as a substrate for bioconversion. An initial glucose concentration was optimized by computer simulation in order to maximize the volumetric productivity. The experiments performed to verify the computer simulation results showed a good agreement with the estimated values which were 1.4 times higher in xylitol yield and 1.85 times higher in volumetric productivity compared with those of the experiments done without glucose under the same conditions. The optimized two-substrate batch fermentation resulted in xylitol yield of 0.81 g-xylitol/g-xylose and xylitol productivity of 5.06 g-xylitol/L·hr. A yield of 0.81g-xylitol/g-xylose is equivalent to 90 % of the theoretical xylitol yield from xylose.

Cell-recycle fermentations were attempted in the second investigation in order to improve xylitol productivity and final xylitol concentration without sacrificing xylitol yield. A series of cell-recycle experiments showed that the feeding of xylose, glucose and yeast extract in the xylitol production phase was the most effective in enhancing xylitol productivity. The optimized cell recycle fermentation resulted in 0.82 g-xylitol/g-xylose yield, 4.94 g-xylitol/L·hr productivity, and 189 g/L final xylitol concentration. Even though cell-recycle fermentations showed very promising experiment results, it may be very difficult to implement cell-recycle fermentation in an industrial scale.

Finally, in the third investigation, the xylose reductase gene from *P. stipitis* was expressed in recombinant *S. cerevisiae* EH13.15:pY2XR, which in turn was able to convert xylose to xylitol very efficiently with a yield of 0.95 g-xylitol/g-

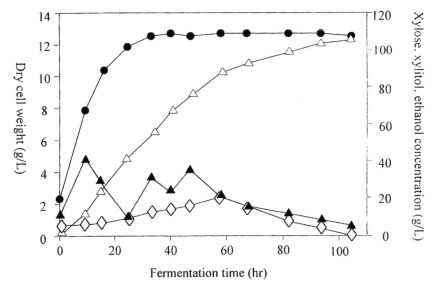

Figure 5. Fed-batch fermentation of S. cerevisiae EH13.15:pY2XR grown in O'Connor's medium at 30 ℃. Glucose concentration was maintained at 0.35 g/L throughout the fermentation.

(● *Dry cell weight (g/L),* ▲ *xylose (g/L),* △ *xylitol (g/L),* ◊ *ethanol (g/L))*

(Reproduced with permission from reference 22. Copyright 2000 ELSEVIER)

Table 1. Comparison of fermentation properties of *C. tropicalis* and recombinant *S. cerevisiae*

	Operation mode	Xylitol yield (g-xylitol/g-xylose)	Xylitol productivity (g-xylitol/L·hr)	Final xylitol concentration (g-xylitol/L)
C. tropicalis	Batch	0.81	5.06	100
C. tropicalis	Cell-Recycle	0.80	4.90	189
Recombinant *S. cerevisiae*	Fed-Batch	0.95	1.69	105

xylose in the presence of glucose used as a co-substrate. However, the supplemented glucose inhibited xylose uptake and caused accumulation of ethanol. Glucose-limited fed-batch fermentations were adopted to solve these problems. By maintaining a high molar ratio of xylose to glucose during the bioconversion phase, 105 g/L xylitol was produced with productivity of 1.69 g/L·hr in O'Connor's medium at 30 °C. The recombinant *S. cerevisiae* strain presented a high xylitol yield, but the productivity was lower than that of natural xylose-fermenting yeasts such as *C. tropicalis* as shown in the first and second investigation. The experimental results in the first, second, and third investigation are summarized in Table 1.

Acknowledgments

This work was supported by research grants from Bolak Company and Ministry of Education through the BK21 program.

References

1. Washüttl, J.; Riederer, P.; Baucher, E. *J. Food Sci.* **1973**, 38, 1262-1263.
2. Deis, R. *Food Technol.* **1993**, 94.
3. Pepper, T.; Olinger, PM. *Food Technol.* **1988**, 10, 98-106.
4. Meakinen, KK. *Adv. Food Res.* **1979**, 25, 137-158.
5. Ylikahri, R. *Adv. Food Res.* **1979**, 25, 159-180.
6. Hyoenen, L.; Koivistoninen, P.; Voirol, H. *Adv. Food Res.* **1983**, 28, 373-403.
7. Vongsuvanlert ,V.; Tani, Y. *J. Ferment. Bioemg.* **1989**, 67, 35-39.
8. Meyrial, V.; Delgenes, JP.; Molletta, R.; Navarro, JM. *Biotech. Lett.* **1991**, 13, 281-286.
9. Kim, SY.; Kim, JH.; Oh, DK. *J. Ferment. Bioeng.* **1997**, 83, 267-270.
10. Gong, CH.; Chen, LF.; Tsao, GT. *Biotech. Lett.* **1981**, 3, 130-135.
11. Oh, DK.; Kim, SY. *Kor. J. Appl. Microbiol. Biotech.* **1997**, 25, 197-202.
12. Yahash, Y.; Horitsu, H.; Kawai, K.; Suzuki, T.; Takamizawa, K. *J. Ferment. Bioeng.* **1996**, 81, 148-152.
13. Lee, H.; Atkin, AL.; Barbosa, MFS.; Dorshied, DR.; Schneider, H. *Enz. Microb. Tech.* **1988**, 110, 81-84.

68

14. Yahashi, Y.; Hatsu, M.; Horitsu, H.; Kawai, K.; Suzuki, T.; Takamizawa, K. *Biotech. Lett.* **1996**, 18, 1395-1400.
15. Audet, P.; Lacroix, C.; Paquin, C. *Thermophilicus. Int. Dairy J,* **1991**, 1, 1-15.
16. Hallborn, J.; Walfridsson, M.; Airaksinen, U.; Ojamo, H.; Hahn-Hagerdal, B.; Penttila, M.; Keranen, S. *Bio/Technol.* **1991**, 9, 1090-1095.
17. O'Connor, GM.; Sanchez-Riera, F.; Cooney, CL. *Biotech. Bioeng.* **1992**, 39, 293-304.
18. Bruinenberg, PM.; van Dijken, JP.; Scheffers, WA. *J. Gen. Microb.* **1983**, 129, 965-971.
19. Evans, TC.; Mackler, B.; Grace, R. *Arch. Biochem. Biophys.* **1985**, 243, 492-503.
20. Kim, JH.; Ryu, YW.; Seo, JH. *J. Ind. Microbiol. Biotechnol.* **1999**, 22, 181-186.
21. Choi, JH.; Moon, KH.; Ryu, YW.; Seo, JH. *Biotechnol. Lett.* **2000**, 22, 1625-11628.
22. Lee, WJ.; Ryu, YW.; Seo, JH. *Process Biochem.* **2000**, 35, 1199-1203.
23. Cillilo, V P. *J. Gen. Bacteriol.* **1968**, 95, 603-611.
24. Kotyk, A. *Fol. Microbiol.* **1967**, 12, 121-131.

Chapter 6

Plate and Disk Bioreactors for Making Bacterial Cellulose

Peter Gostomski[1], Henry Bungay[2], and Richard Mormino[2]

[1]Department of Chemical and Process Engineering, University
of Canterbury, Christchurch, New Zealand
email: p.gostomski@cape.canterbury.ac.nz
[2]H. P. Isermann Department of Chemical Engineering, Rensselaer
Polytechnic Institute, Troy, NY 12180–3590 email: bungah@rpi.edu or
mormir@rpi.edu

The product from manufacturing bacterial cellulose can be a
tough gel (pellicle), reticulated filaments, or pellets.
Inefficient and technologically crude large scale surface
culture has been successful in the Republic of the Philippines
because of vigorous growth and lowering of pH to discourage
potential contaminating organisms. We observe superior
performance by supporting the organisms with plates or disks
and alternately wetting them with nutrient media and exposing
them to air. Power consumption is low, and no expensive
system is need for aeration. Reactor configurations suitable for
low technology agribusiness have been evaluated.

© 2002 American Chemical Society

69

Introduction

Bacterial cellulose is used in foods, in acoustic diaphragms for audio speakers or headphones, for making unusually strong paper, and has medical applications such as wound dressings and artificial skin (1-9). Most of the bacterial cellulose that is currently an article of commerce comes from agitated, deep culture fermentation with strains of *Acetobacter xylinium* that form no pellicle but produce reticulated fibers (10). Pellets made in an air-lift fermenter have been introduced recently that have properties that fall in between the pellicle from surface cultures and the fibrous version (11).

In static cultures, *A. xylinium*, a gram-negative, aerobic, rod-shaped organism, produces cellulose at the air-water interface as an assembly of highly crystalline interwoven ribbons that are chemically pure, free of lignin and hemicellulose, and have a high degree of polymerization (12). The thick, tough pellicle, in its undried state, has outstanding hydrophilicity. Depending on synthesis conditions, it may have a water holding capacity ranging from 60 to 700 times its dry weight. Under suitable conditions, as much as 50% of the supplied carbon substrate may be assimilated into cellulose from a wide variety of sugars.

Food applications and some medical uses require the pellicular form of bacterial cellulose. However, the pellicle can be disintegrated to get very tough fibers very similar to those from the reticulated form. Most of the water can be squeezed from the pellicle, and little or no further drying would be required for saleable fibers. Pellicular bacterial cellulose presents a good opportunity for low-investment agribusiness because of production with little investment of heat or mechanical energy and easy product recovery. Commercialization should be straightforward in view of success of many small companies in the Republic of the Philippines with a pellicular product, Nata de coco, that had peak sales of $28 M in 1993, mostly for food consumption as a dessert material.

Static surface culture is slow and requires much labor. Disk and plate bioreactors have higher volumetric efficiency and may be operated continuously. The rotating biological contactor (RBC), widely used for treatment of waste water, has desirable features of high area for attachment, low power consumption, continuous operation, and easy scale-up (13). Our bioreactors for bacterial cellulose were inspired by treatment of wastes but have an important difference in that perforated or mesh disks retain the organisms in the growing product whereas the reactors for waste treatment encourage sloughing of cells for disposal. A disk bioreactor outperforms surface cultures grown in trays. While surface culture runs last 10 to 14 days, the disk bioreactor produces similar bacterial cellulose in less than 4 days but with higher water content (14). Various designs that correct drawbacks in the rotating disks are presented here.

Reactor Construction

The rotating disc reactor consists of a cylindrical trough that contains the medium, a shaft holding the rotating disks, and a motor connected to the shaft (14,15). The trough is made of a transparent polymeric material (polycarbonate or acrylic) to allow close observation of film growth. The disk bioreactor is 30 cm long and 14 cm in diameter, with two separate bottom compartments each having a working volume of 1 liter. Sometimes disks are omitted on one side to provide a static control for the rotating disks on the other side. Flat acrylic plates are used to seal the two sides of the cylinder. The cylinder is cut in half and connected with hinges for opening the bioreactor.

Circular plastic meshes commonly used as patterns for cross stitching are a good material for making disks (as few as one or as many as a dozen) on a shaft that turns so that they dip into the nutrient medium, continue up into the air, and dip again. Soon some submerged fibrils of bacteria attach to the disks and grow and form cellulose as a gel that increases in thickness.

Although the rotating disk bioreactor performs well, the final product from a disk is circular with a hole in its center for the shaft. A large circular sheet from a commercial bioreactor would suffer losses as squares or rectangles were cut from it, and there would be additional waste because of the central hole. A variety of new designs address this problem.

Some improvement over circular disks comes from replacing the central shaft with a plastic cylinder with slots through which the ears of square elements of fabric protrude. Ears of each element are bent back and held

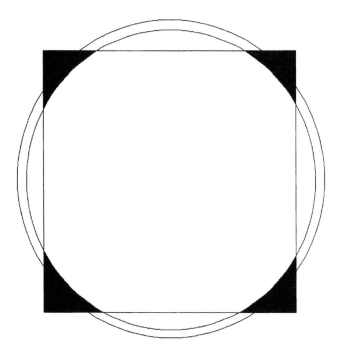

Figure 1. Square element of fabric in a rotating cylinder. The dark regions are the ears that are essentially unproductive.

in place by a large rubber band around the plastic cylinder. Blocks of plastic are cemented at each end of the cylinder to connect a shaft for rotation. Figure 1 is a sketch of one element.

This bioreactor was tested with coarse cheesecloth and with cloth cut from a muslin cotton bed sheet. The cheesecloth drained too rapidly and had poor formation of cellulosic gel near the center where it could not dip into the nutrient medium because the shaft rotated slightly above the liquid level. The cloth cut from the bed sheet worked well because capillary action distributed liquid better.

The next disadvantage that was addressed was volumetric efficiency. A room full of rotating bioreactors would sacrifice space because circles cannot fill a space no matter how they are packed. There are alternative ways to accomplish the basic concept of alternate aeration and liquid nutrition with a rectangular geometry that can fill almost all of the available space. In fact, our very first reactor design was a hanging sheet to which liquid medium was applied by dripping. This design was abandoned because of poor and irreproducible distribution of the liquid. A new design inclines the mesh. This slows the flow, allows more time for liquid to spread over the mesh, and reduces the distance from top to bottom.

An array of rectangular elements is constructed from two racks intended for elevating a stack of papers above other items on a desk. These racks consist of small circular rods coated with paint and are 14 cm high, 40 cm long, and 21 cm wide. Racks about half as large can also be purchased. The legs are bent to the desired angle (60 to 75 degrees) and cut to the desired length. One rack is inverted, and the legs are joined to the legs of the other rack with rubber tubing. Fabric is threaded over and under the top and bottom rods to form the sheets for growing the cellulose as shown in Figure 2. Not counting the portion between panels, each fabric panel for these tests is 27 cm long and 20 cm wide, inclined at 70 degrees. There are 21 bars at the top and at the bottom, but some are unused such that the spacing of the sheets is 4 cm.

Each design with inclined, rectangular elements has an enclosed array of inclined mesh sheets inside of a trough for the nutrient medium. If a method for continuous harvesting of the pellicle cannot be devised, a fresh array could substitute for one that was removed for harvesting.

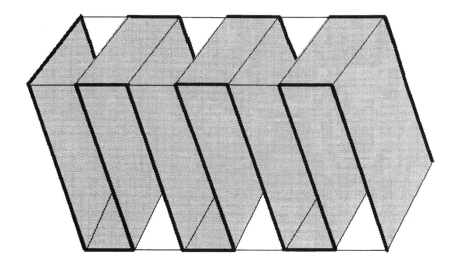

Figure 2 Elements formed by threading fabric around rods

Alternatives for providing nutrient medium to cells adhering to a mesh are immersion and simple down flow. Immersion can be by bringing the medium to the elements or by dipping the elements into the medium. Immersion by pumping is accomplished by transferring medium between the reactor and a reservoir. This is a simple and effective design for research. However, moving the medium from one vessel to another with a cycle time of 12 changes per minute (based on a good rpm found for the rotating disk bioreactor) demands extreme flow rates and concomitant high energy consumption. However, this design permits the submerged time to be different from the aeration time in contrast to the rotating disks in which aeration and submerged times are roughly equal.

Dipping the mesh elements into the medium should use two matched mesh assemblies that are raised and lowered into the medium in a seesaw arrangement so that one is the counterweight for the other. The concept is shown in Figure 3 but has not been tested yet.

The down flow design pumps medium to the top, collects the drainage, and pumps it back to the mesh. Liquid distribution benefits from application of medium to the horizontal cloth between elements. Attention is now focussed on equalizing the amount of liquid to each element.

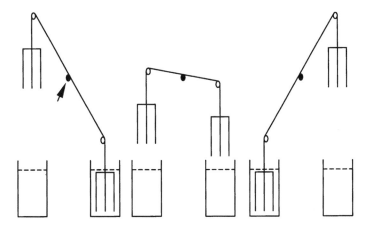

Figure 3. Dipping of elements into nutrient medium

76

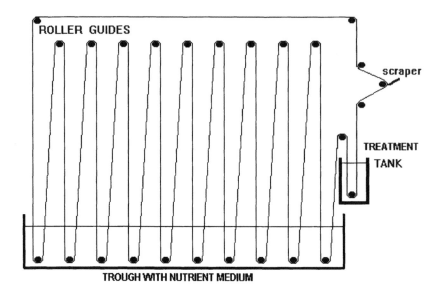

Figure 4. Continuous belt bioreactor

Another design that makes much sense for commercialization is too complicated for testing at present. It uses a continuous mesh belt that dips into troughs of medium, reaches a scraper that removes the product, and the belt continues back for inoculation and growth. A sketch of the arrangement is Figure 4. A simplified version of this design has been reported for producing continuous filaments (16).

Removal of cells

Bacteria that produced the cellulosic gel are retained in the product. One common practice is to hydrolyze the bacteria with hot caustic and to flush away the debris by soaking in water. This has the potential for hazardous operations and generation of waste waters if practiced on a large scale. For some applications, the organisms might be permitted in the gel or in dried sheets made from it. An example is rough cardboard in which dead bacteria would be tolerable. Our current procedure for removal of cells is to add 2 or 3 drops of toluene to several liters of water in which the gel is being soaked. Autolysis is induced, and cell debris is flushed out during soaking in changes of tap water. This removal should be possible in the bioreactor before the pellicle is harvested.

Conclusion

All of the mesh systems that have been tested form cellulosic gel at higher rates than observed with surface cultures in trays. Performance of rectangular arrays is similar to that with rotating disks, and the remaining engineering problems for distribution of liquid, simple harvesting of product, and process control do not seem to be formidable.

Acknowledgements

This research was started with Grant BES-9501809 from the National Science Foundation. It continued with Grant 99-35504-8011 from the US Department of Agriculture. Bernardo da Costa constructed the reactor systems with rectangular plates and refined the designs.

References

1. White, D. G.; Brown, Jr. R. M. In *Cellulose and Wood-Chemistry and Technology;* Schuerch, C., Ed.; John Wiley & Sons: New York, 1989; pp 573-590.
2. Yamanaka, S.; Watanabe, K.; Kitamura, N.; Iguchi, M.; Mitsuhashi, S.; Nishi, Y.; Uryu, M. *Jour. Material Sci.* **1989,** *24,* 3141-3145.
3. Brown, Jr. R. M. In *Cellulose: Structural and Functional Aspects;* Kennedy; Phillips; Williams., Eds.; Ellis Horwood Ltd., 1989; pp 145-151.
4. Brown, Jr. R. M. In *Polymers from Biobased Materials;* Chum, H.L., Ed.; Noyes Data Corp.: Park Ridge, NJ, 1991.

78

5. Brown, Jr. R. M. In *Harnessing Biotechnology for the 21ˢᵗ Century;* Ladisch, M. R.; Bose, A., Eds.; American Chemical Society, 1992; pp 76-79.

6. Okiyama, A.; Motoki, M.; Yamanaka, S. *Food Hydrocolloids.* **1993**, *6,* 503-511.

7. Yamanaka, S.; Watanabe, K. In *Cellulose Polymers, Blends, and Composites;* Gilbert, R. D., Ed.; Hanser/Gardner Publications: Cincinnati, OH, 1994; pp 207-215.

8. Geyer, U.; Heinze, T.; Stein, A.; Klemm, D.; Marsch, S.; Schumann, D.; Schmauder, H. P. *Intl. Jour. Biol. Macromol.* **1994**, *16(6),* 343-347.

9. Takai, M. In *Cellulose Polymers, Blends, and Composites;* Gilbert, R. D., Ed.; Hanser/Gardner Publications: Cincinnati, OH, 1994, pp 233-240.

10. Cannon, R. E.; Anderson, S. M. *Microbiol.* **1991**, *17,* 435-447.

11. Johnson, D. C.; Neogi, A. N. U.S. Patent 4,863,565, 1989.

12. Chao, Y.; Ishida, T.; Sugano, Y.; Shoda, M. *Biotechnol. Bioeng.* **2000**, *68(3),* 345-352

13. Grady, Jr. C. P. L; Lim, H. C. In *Biological Wastewater Treatment: Theory and Application;* Marcel Dekker, Inc.: New York, 1980.

14. Bungay, H. R.; Serafica, G. C. U.S. Patent 5,955,326, 1999.

15. Bungay, H. R.; Serafica, G. C. U.S. Patent 6,071,727, 2000.

16. Sakairi, N.; Asano, H.; Ogawa, M.; Nishi, N.; Tokura, S. *Carbohydrate Polymers* **1998**, *35,* 233-237.

Chapter 7

The Effect of the Dissolved Oxygen Concentration on the Production of Lignin Peroxidase and Manganese Peroxidase by *Phanerochaete chrysosporium*

Sue Hyung Choi, Man Bock Gu*, and Seung-Hyeon Moon

Department of Environmental Science and Engineering, Kwangju Institute of Science and Technology (K-JIST), 1 Oryong-dong, Puk-gu, Kwangju 500–712, South Korea
*Corresponding author: email: mbgu@kjist.ac.kr

The effect of the dissolved oxygen (D.O.) concentration on the production of lignolytic enzymes and isozyme distributions by *Phanerochaete chrysosporium* was studied in the immobilized reactor system. The oxygen levels significantly affected the production of lignin peroxidase (LiP) and manganese peroxidase (MnP), as well as that of H_2O_2. It is known that a high oxygen level is required to produce these enzymes. In this study, however, D.O. concentrations above a critical D.O. concentration (1.5 fold increase over the saturated concentration with air) inhibited LiP and MnP productions in the experiments when oxygen was continuously supplied. It is thought that a greater H_2O_2 accumulation seen with higher D.O. concentrations caused adverse effects on the LiP and MnP productions. On the other hand, with lower D.O. concentrations, H_2O_2 was not produced enough to stimulate either LiP or MnP production. To reduce excess H_2O_2 accumulation during culturing with a continuous supply of oxygen, a pulsed supply of oxygen was introduced.

© 2002 American Chemical Society

INTRODUCTION

Phanerochaete chrysosporium establishes the lignin-degrading system through the secretion of secondary metabolites, including two isozyme families of secreted peroxidases, lignin peroxidase (LiP) and manganese peroxidase (MnP), as well as oxalate, veratryl alcohol, and H_2O_2 [1, 2]. The substrate specificity of this lignin-degrading system is so low that many aromatic compounds, which have a similar structure to lignin, can be degraded by this system [3-7].
Various culture parameters for enhanced production of MnP and LiP have been studied [1, 2, 4]. Among them, oxygen has been found to be a critical factor determining the production level of these enzymes [8, 9]. To date, most laboratory-scale studies of ligninase production in submerged or stationary cultures of *P.chrysosporium* have used periodic purging with pure O_2 [10, 11]. Recently, many different methods of supplying oxygen to increase the MnP and LiP productions have been studied. These methods were mainly studied in the cultures and simply differed by their methods of supplying of oxygen, including continuous or periodically-pulsed methods [2, 12]. In these studies, though, the dissolved oxygen concentration can not be controlled while the pH would also change dependent upon the culture time.
Therefore, to investigate the role of the oxygen tension in the production of MnPs and LiPs, a more systematic experiment, in which controls are added to measure and maintain the dissolved oxygen concentration as well as the pH, is needed. Therefore, in this study, we used a bioreactor, which can control both the pH and D.O., to manage the oxygen concentration as the only changeable parameter, and found that oxygen had a positive as well as negative effect on the LiP and MnP productions. To illuminate the possible mechanism of this phenomenon, the H_2O_2 concentration was measured and a possible reason was suggested.

MATERIALS AND METHODS

Strain and Medium Compositions

Phanerochaete chrysosporium ATCC 24725 was maintained at 39°C on malt-agar plates. Spores were harvested and filtered through glass wool. The spore concentration was determined by measuring the absorbance level at 650 nm (an absorbance of 1.0 cm^{-1} is approximately 5×10^6 spores/mL). The growth medium was the same as Tien and Kirk used [13], except that veratryl alcohol and Tween 80 were added after two days of growth.

Cell Immobilization on Polyurethane Foam

The supporting matrix for immobilization was polyurethane foam, synthesized in the Department of New Materials Science and Engineering at the Kwangju Institute of Science and Technology (K-JIST) [3-5]. A total of 15 grams of foam sliced into cubes (5 mm × 5 mm × 5 mm) were placed in a 1.5 L of the reactor and autoclaved at 121 °C for 20 min. The spore concentration in the reactor was adjusted to 6×10^{5} spores / mL in a total volume of 800 mL.

Reactor operation

The reactor was operated with conditions set at pH 4.5, a temperature of 39°C, and a rotor speed of 120 rpm. Every condition, except the D.O. concentration, was maintained constant in all experiments. Nitrogen and air were used to adjust the D.O. concentrations between 0% and 100 %, i.e. a value equivalent to the D.O. concentration resulting from bubbling sterile atmospheric air into the bioreactor. For values above 100 % D.O., pure oxygen was used. To maintain a constant D.O. concentration, the auto-regulator of the Bioflo3000 bioreactor was used.

Measurements of Enzyme Activities and SDS-PAGE

Lignin peroxidase activity was measured using veratryl alcohol [13] and MnP activity was measured with Mn(II) as the substrate according to the methods described previously [14]. Equal volumes of extracellular fluid were concentrated 25-fold using membrane filters, Microcon-30 (Amicon Co., USA), and subjected to sodium-dodecyl sulfate - polyacrylamide gel electrophoresis (SDS-PAGE) using a 3% stacking gel and a 12 % running gel. Proteins were visualized by staining with commassie brilliant blue.

Measurement of Glucose and H_2O_2

The glucose concentration in the media was measured using a glucose-kit (Young-dong co., Korea) while the H_2O_2 concentration was determined following the method of Ziang et al. [15].

RESULTS AND DISCUSSIONS

Continuous, constant D.O. control

Using a continuous supply of oxygen to maintain a constant D.O. level, five different D.O. concentrations, 70 %, 100 %, 150 %, 260 %, and 345 %, were tested and compared in terms of the culture's LiP activity, MnP activity, and H_2O_2 production. When only air was used to 70 % D.O., which was below the saturated concentration seen with air, the glucose was consumed over a 25 day period but no LiP activity, MnP activity or H_2O_2 was detected (Figure 1). Therefore, it would seem that the D.O. concentrations below air-saturated conditions were not sufficient to result in LiP or MnP activities as well as H_2O_2 production, regarded as an important factor leading to induction of LiP's and MnP's. On the other hand, when the D.O. was at or above the air-saturated concentration, i.e. 100 % D.O., 150 % D.O., 260 % D.O. and 345 % D.O., LiP and/or MnP activities and H_2O_2 production were detected. But, unexpectedly, as the D.O. concentration increased, the MnP activity decreased, while LiP activity was only detected with 100 % D.O. (Figure 2).

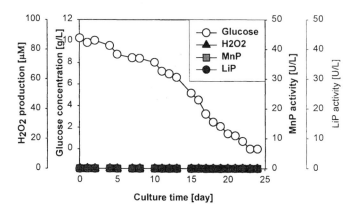

Figure 1. Time course of ligninase activities and H_2O_2 production with 70 % D.O. concentration, which is a below the saturated concentration with air

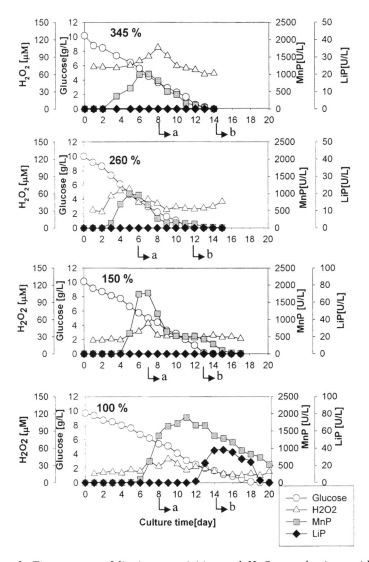

Figure 2. Time course of ligninase activities and H2O2 productions with a continuous supply of oxygen above the saturated concentration level achieved with air

Specific production patterns of LiP and MnP dependent on the D.O. concentration

As D.O. concentration increased, the period for complete glucose consumption was shortened while the H_2O_2 production increased (Figure 2). In addition, H_2O_2 accumulated early on the culture and then decreased after the maximum MnP production was reached. Hence, H_2O_2 production followed a bell-shaped production curve. The maximum value of H_2O_2 observed in each of the experiments increased in cultures having a higher oxygen concentration (Figure 2).

The initiation of MnP production was also faster with higher oxygen concentrations, while the maximum MnP activity decreased with increasing oxygen concentrations. It would appear that the specific LiP and MnP production patterns correlated with the H_2O_2 production, which was further dependent upon the D.O. concentration. Comparing the H_2O_2 production and MnP activity, it was shown that the MnP activity sharply decreased after the maximum H_2O_2 production. From these results, the best D.O. concentration for MnP activity and LiP activity was selected as 100 % D.O., which gave a plateau-shaped H_2O_2 production profile.

Therefore, it is thought that the H_2O_2 that accumulated might cause the adverse effects seen with the LiP and MnP activities. There are two possible ways for H_2O_2 to cause such adverse effect. Firstly, excess H_2O_2 can result in enzyme inactivation [2]. Therefore, even though the enzymes were produced, they do not react with the substrate and shows no enzyme activity. Secondly, the enzyme is not produced primarily through inhibition or damage to the cell brought on by the excess H_2O_2, which is referred to as a production-level inhibition. To elucidate the possible relationship between H_2O_2 accumulation and LiP and MnP activities, enzyme levels were analyzed using SDS-PAGE analysis. SDS-PAGE data showed that for both LiP and MnP, their activities (Figure 2) correlated well with the protein levels (Figure 3). When cultures were grown with 350 % D.O. and 250 % D.O., which were a 3.5-fold increase and 2.6-fold increase over the saturated concentration with air, respectively, MnP activity totally disappeared 7 days after reducing a maximum (Figure 2) This result was also seen in the SDS-PAGE data (Figure 3, lane 1-4). On the other hand, with 150 % D.O., the MnP activity was maintained over a relatively longer period as seen with both enzyme assays (Figure 2) and SDS-PAGE gel (Figure 3, lane 6,7). As can be seen in activity data (Figure 2), both LiP and MnP activities were detected when using 100 % D.O. while the activity of MnP was stably maintained for a longer time. Again, the enzymes patterns seen on the SDS-PAGE gel correlated well with the activity data (Figure 3, lane 7,8). Therefore, the abrupt decrease in MnP activity after accumulation of H_2O_2 and no detection of LiP activity, except with 100 % D.O., were mainly the results of lower production according to the SDS-PAGE data, although there might be some enzyme inactivation due to excess H_2O_2.

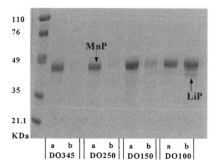

Figure 3. SDS-PAGE showing the isozyme distribution when a continuous supply of oxygen above the saturated concentration with air was provided. a and b are two samples that correspond to the arrows in Figure 2.

Pulsed Supply of Oxygen

As described in the previous sections, an accumulation of H_2O_2 during culturing with a continuous supply of oxygen resulted in the inhibition of LiP and MnP productions. Thus, to reduce H_2O_2 accumulation during culture periods, a pulsed supply of oxygen was introduced instead of a continuous supply of oxygen.

As expected, changes in the way that oxygen was supplied prevented a bell-shape H_2O_2 accumulation, but showed a constant level production, a plateau. In addition, MnP activity was maintained for longer periods than in the continuous supply experiments (compare Figure 2 and Figure 4) while LiP activity was seen in all the pulsed experiments. These LiP and MnP activity trends (Figure 4) correlated well with their production (Figure 5). From the SDS-PAGE gel (Figure 5), MnP production was stably maintained for a longer period of time, while LiP production occurred in all experiments, regardless of the oxygen concentration.

However, higher D.O. concentrations during the pulsed experiments were still not good for MnP nor LiP production. Figure 6 compares the time course of the enzyme's activities and their production for both the continuous oxygen supply experiments and pulsed oxygen supply experiments. This figure clearly demonstrates that the pulsed case is better than the continuous supply in terms of MnP and LiP production. Table 1 summarizes all distinguishable patterns compared between the continuous oxygen supply and pulsed oxygen supply experiments.

86

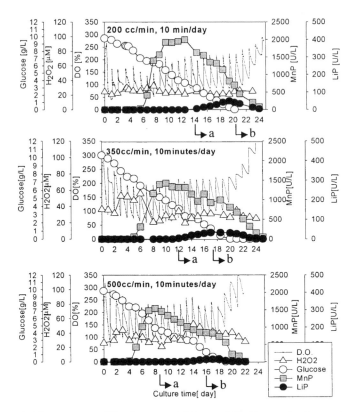

Figure 4. Time course of ligninase activities and H_2O_2 productions when pulse oxygen was added.

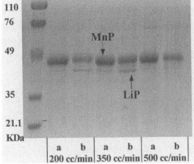

Figure 5. SDS-PAGE showing the isozyme distribution for experiments having a pulsed suppy of oxygen. a and b are two samples that correspond to the arrows in Figure 4.

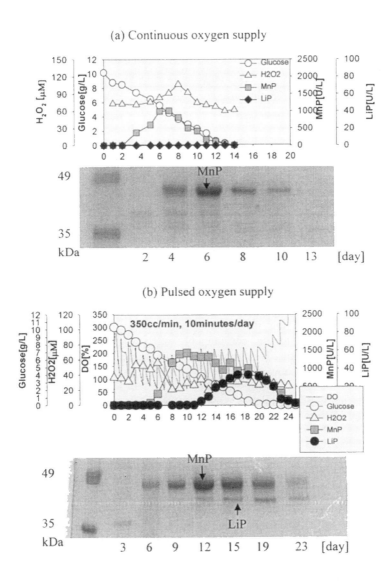

Figure 6. The comparison of ligninase activities and production between the continuous supply of oxygen and the pulsed supply of oxygen

Table 1. Summary of culture characteristics according to the oxygen supplying method

	As the D.O. concentration increased (above the saturated concentration with air)	
	Continuous Supply of O_2	Pulsed Supply of O_2
Glucose consumption rate	Increased	Increased
Start-up time for MnP production	Faster	Faster
MnP (maximum)	Decreased	Slightly decreased
LiP production	Occurs only at 100 % D.O.	Occurs for all cases
H_2O_2 production	Increased (Shows Max. value/Bell curve)	Slightly increased (Plateau)
MnP acitivity	Sharply decreased (After H_2O_2 peak)	Sustained longer

On the other hand, when the reactor system was operated without any pH control, the pH during the culture was between 4.7-5.5. During these experiments, the glucose consumption rate, MnP production, and LiP production all decreased compared with the value seen when the pH was controlled (data not shown).

The relation between the D.O., the reactive oxygen species, and the enzyme activities

The optimum pH for a *P. chrysosporium* culture is 4.5 [16] and it is known that O_2^- can easily be converted to H_2O_2 by obtaining one electron from the cell at this pH [17]. Other methods to form H_2O_2 by the cells includes enzymatic reactions, such as glucose oxidase and NADPH oxidase [18, 19]. With these enzymes, H_2O_2 can be produced directly from molecular oxygen.

In this study, it was found that as the D.O. concentration increased, H_2O_2 production also increased. This is because an increase in the D.O. concentration accelerates H_2O_2 production through the methods described above. On the other hand, the H_2O_2 concentrations suddenly show a decrease, at which time it is thought that H_2O_2 might be converted to the OH^- which is more toxic than H_2O_2. One possible reason for the production of OH^- is the presence of Fe^{2+} and Mn^{2+} in the media, which reacts with H_2O_2 to form OH^- through the Fenton reaction. It is also known that the white rot fungi, *Irpex lacteus* and *Phanerochaete chrysosporium,* secrete a low-molecular-weight substance containing Fe^{2+} that catalyzes a redox reaction between electron donors and O_2, to produce the hydroxyl radical via H_2O_2 [20].

Comparing the reactive oxygen species in terms of their toxic effect, firstly O_2^{-}, which is formed from oxygen through a one electron transfer, is known to be potentially toxic, and this can be converted to H_2O_2 by obtaining one election and two protons. Finally, OH^{-}, which is the most toxic form, can be produced from H_2O_2 [21]. H_2O_2 and OH^{-} cause damage to the cell's DNA and the proteins [21], and so it is possible to affect LiP and MnP productions with these oxygen species. In addition, the abrupt decrease in MnP activity after a peak level of H_2O_2 production seems to be the result of inhibition brought on by the production of the hydroxyl radical.

It was reported that when exogeneous H_2O_2 was added to a 5-day-old, nitrogen limited, Mn-deficient cultures of *P.chrysosporium* to a final concentration between 0.25 and 8 mM, accumulation of *mnp* mRNA was detected in cells exposed to H_2O_2 concentrations between 0.25 to 1.5 mM, while no *mnp* mRNA was detected in the absence of H_2O_2 or with H_2O_2 concentration above 1.5 mM [22]. These facts support our results that oxygen was needed to enhance the MnP and LiP productions, but that excess dissolved oxygen may cause inhibition in both their productions due to the generation of reactive oxygen species, including H_2O_2 and the hydroxyl radical.

As a result, a pulsed oxygen supply was better than a continuous oxygen supply in terms of LiP and MnP productions. As well, it is thought that there might be a critical D.O. concentration for the enhancement of LiP and MnP productions since H_2O_2 was needed for the stimulation of their productions while higher values of H_2O_2 might inhibit their production.

CONCLUSIONS

Oxygen is needed to enhance the MnP & LiP productions, but excess dissolved oxygen appears to result in an inhibition in LiP & MnP productions due to the generation of excess reactive oxygen species. The use of a pulsed oxygen supply was better than a continuous oxygen supply in terms of LiP and MnP productions because H_2O_2 production was maintained efficiently in a pulsed oxygen supply without showing much accumulation.

ACKNOWLEDGEMENT

This research was funded by a K-JIST independent project (1999-2000) and in part by the Korea Science and Engineering Foundation (KOSEF) through the ADEMRC at Kwangju Institute of Science and Technology (K-JIST).

90

REFERENCES

[1] Orth, A. B.; Pease, E. A.; Tien, M. In *Biological degradation and bioremediation of toxic chemicals*; Chaudhry, G. R., Ed.; Discorides Press: USA. 1994; pp345-363.

[2] Buswell, J. A.; Odier, E. *CRC Crit. Rev. Biotechnol.* 1987, *6*, 1-60.

[3] Choi, S. H.; Moon, S.-H.; Lee, J.-S.; Gu, M. B. *J. Microbiol. Biotechnol.* 2000, *10*, 759-763.

[4] Choi, S. H.; Song, E.; Gu, M. B; Moon, S.-H. *Korean J. Biotechnol. Bioeng.* 1998, *13*, 223-230.

[5] Choi, S. H.; Song, E.; Lee, K. W.; Moon, S.-H.; Gu, M. B. In *Bioremediation and Phytoremediation;* Wickramanayake G. B. and Hinchee R. B., Eds.; The first international conference on Remediation of Chlorinated and Recalcitrant Compounds; Battle press: Columbus, USA, 1998, pp161-166.

[6] Chung, N.; Aust, S. D. *Arch. Biochem. Biophys.* 1995, *316*, 851-855.

[7] Chung, N.; Aust, S. D. *Arch. Biochem. Biophys.* 1995, *322*, 143-148.

[8] Faison, B. D.; Kirk, T. K. *Appl. Environ. Microbiol.* 1985, *49*, 299-304.

[9] Kirk, T. K.; Farrell, R. L., Annu. Rev. Microbiol. 1987, *41*, 465-505

[10] Leisola, M. S. A.; Fiechter, A. *FEMS Microbiol. Lett,* 1985, *29*, 33-36

[11] Jäger, A.; Cronan, S.; Kirk, T. K. *Appl. Environ. Microbiol.* 1985, *50*, 1274-1278.

[12] Bonnarme, P.; Delattre, M.; drouet, H.; Corrieu, G; Asther, M. *Biotechnol. Bioeng.* 1993, *41*, 440-450.

[13] Tien, M.; Kirk, T. K. *Methods Enzymol.* 1988, *161*, 238-249.

[14] Gold, M. H.; Glenin, J. K. *Methods Enzymol.* 1988, *161*, 258-264.

[15] Jiang, Z.-Y., Hunt, J. V., Wolff, S. P. *Anal. Biochem.* 1992, *202*, 384-389.[16] Tien, M. *Critical Reviews in Microbiology* 1987, *14*, 141-168.

[17] Bielski, B. H. J., Cabelli, D. E., Arudi, R. L., Ross A. B. *J. Phys. Chem. Ref. Data.* 1985, *14*, 1041-1051.

[18] Asada, Y.; Miyabe, M.; Kikkawa, M.; Kuwahara, M. *Agric.Biol. Chem.* 1986, *50*. 525-259

[19] Kelly, R. L.; Reddy, C. A. *Methods Enzymol.* 1998, *161B*, 307-316.

[20] Tanaka, H.; Itakura, S.; Enoki, A. *J. Biotechnol.* 1999, *75*, 57-70.

[21] Biology of Microorganisms; Brock, T. D.,; Madigan, M. T. Eds.; 6th edition; Prentice-Hall International, Inc.: Canada, USA. 1991. pp.331-332.

[22] Li, D., Alic, M., Brown, J. A., Gold, M. H. *Appl. Environ. Microbiol.* 1995, *61*, 341-345.

Chapter 8

Development of Effective Immobilization Method Using Modified Glutaraldehyde for GL-7-ACA Acylase

Seung Won Park[1], Sang Yong Choi[2], Koo Hun Chung[2], Suk In Hong[1], and Seung Wook Kim[1,*]

[1]Department of Chemical Engineering, Korea University, 1, Anam-dong, Sungbuk-ku, Seoul 136–701, Korea
[2]Chong Kun Dang Pharmaceutical Corporation, Seoul, Korea

The objectives of this study are to develop a suitable immobilization method for GL-7-ACA acylase, and to enhance the stability of GL-7-ACA acylase immobilized on silica gels. In this study, each step in the immobilization procedure was optimized individually to improve the efficiency of overall immobilization procedure of GL-7-ACA acylase. The optimal conditions of each step were determined as follows; pretreatment by 35% hydrogen peroxide, silanization by 3-APTES in acetone, crosslinking by using modified glutaraldehyde and 10 mg/ml as amount of GL-7-ACA acylase added in immobilization procedure. We also investigated the effect of sodium borohydride in order to increase the stability of the linkage between enzyme and support after immobilization. The activity of immobilized GL-7-ACA acylase treated with 2%(w/w) sodium borohydride was remained about 86% after 20 times of reuse.

© 2002 American Chemical Society

Enzymatic transformation of CPC (Cephalosporin C) into 7-ACA can be performed by a two step process including the oxidative deamination of CPC to GL-7-ACA (glutaryl-7-aminocephalosporanic acid) catalyzed by a D-amino acid oxidase and the subsequent hydrolysis catalyzed by GL-7-ACA acylase(*1-3*). The 7-ACA produced by GL-7-ACA acylase is a useful intermediate of great commercial interest for the preparation of semisynthetic cephalosporin antibiotics. 7-ACA production using microorganism or purified enzyme have many problems that it must be removed from product and cannot be reused.

Immobilization of enzymes has been extensively developed in the industrial applications of enzymes. There are several reasons for the preparation and use of immobilized enzymes. In addition to a more convenient handling of enzyme preparations, the two main targeted benefits are easy separation of the enzyme from the product, and reuse of the enzyme. Inorganic materials have been successfully used for immobilization of enzyme. Silanization to activate supports and subsequent covalent binding of the enzyme to the carrier using a coupling reagent such as a glutaraldehyde is common method used in immobilization of enzyme(*4*).

Silica gel is an amorphous inorganic polymer composed of siloxane groups (Si-O-Si) in the inward region and silanol groups(Si-OH) distributed on the surface. Chemical modifications that can occur with this polymer are related to the presence of the disposed silanol groups in its surface (*5*).

Surface modifications are usually achieved with silanization by using appropriate organosilane agents (*6-8*). Aqueous silanization with a trifucntional silane (aminopropyltriethoxysilane) has been selected since it results in more stable and uniform immobilized enzyme layer (*9*).

Glutaraldehyde has been used as a crosslinker for immobilization of enzyme in which ε -amino groups of lysine in proteins are expected to form a Schiff base with glutaraldehyde. However, the reaction of glutaradehyde with proteins is highly heterogeneous and complicated (*10*). Wilheim and Frank (*11*) suggested that simple Schiff bases are not very stable and the reactivity of enzymes with glutaraldehyde is remarkably low. Terumichi et al.(*12*) have previously reported that the activity of alcohol dehydrogenase increased by using alkaline treated gluataraldehyde, suggesting alkaline treated gluataraldehyde may act as crosslinkers in a manner different from the generally accepted Schiff base formation reaction. A possible mechanism may involve the additional reaction of an amino group to the double bond in the aldol condensate of glutaraldehyde.

Although the literatures have already been reported on the immobilization of GL-7-ACA acylase by different methods such as covalent attachment and ionic binding on organic and inorganic carriers(*13-15*), study on GL-7-ACA acylase immobilized on silica gel has not yet been reported.

In this study, GL-7-ACA acylase was immobilized on silica gel modified by silanization, followed with glutaraldehyde for the production of 7-ACA. The objectives of this study are to develop a suitable immobilization method for GL-7-ACA acylase, and to enhance the stability of GL-7-ACA acylase immobilized on silica gels.

Materials and methods

Materials

Glutaraldehyde was obtained from Fluka Co. ρ -Dimethylaminobenz-aldehyde was obtained from KANTO Chemical Co. Glutaric anhydride was obtained from Aldrich Co. 3-aminopropyltriethoxysilane(3-APTES), 3-aminopropyltrimethoxysilane(3-APTMS) and 3-(trimethoxysilyl) propylethyl-endiamine (3-TMSPEDA) were obtained from Aldrich Co. Silica gel and 7-ACA were supplied by Chong Kun Dang Pharmaceutical Corp. Ultrafiltration membrane (15659-00-1) was purchased from Sartorius.

Preparation of GL-7-ACA acylase

GL-7-ACA acylase from genetically engineered *E. coli* BL21 that contains the GL-7-ACA acylase gene of *Pseudomonas sp. K*AC-1 was supplied by Chong Kun Dang Pharmaceutical Corporation. The preparation procedure of the enzyme was as follows. Ammonium sulfate was added up to 20% saturation in the solution of cell extracts and the suspension was centrifuged (10,000 rpm for 15 min) to obtain the supernatant. Then, precipitate obtained after addition of ammonium sulfate to 40% saturation was resuspended in 100 mM phosphate buffer (pH 8). The solution was dialyzed and concentrated by a ultrafiltration membrane (Cut off MW 30,000). All purification procedures were performed at 4 °C unless stated otherwise. The solution was stored at -20 °C and used for the immobilization.

Synthesis of GL-7-ACA

GL-7-ACA was prepared by the method of Shibuya et al. (*16*). A 15.2 g of glutaric anhydride in 10 mL of acetone was added to a solution, which was prepared by dissolving 9.07 g of 7-ACA in 70 mL of 1.0M sodium bicarbonate (pH 9.0). The reaction mixture was stirred at room temperature for 10 min, and then evaporated under reduced pressure at 25 ℃ for 10 min to remove acetone. The solution was acidified with 5.0 N HCl to pH 1.5 and extracted with 120 mL of ethyl acetate, and quickly filtered with suction. The combined ethyl acetate layers from three such runs were filtered through a Millipore filter(0.45 μm) and concentrated to about 50 mL under reduced pressure at 25 ℃. Chloroform (450 mL) was added to the concentrated solution, and mixed well. The precipitates were collected by filtration and washed thoroughly with 50 mL of ethyl acetate.

Immobilization Method of GL-7-ACA acylase

One gram of dry silica gels was mixed with 3-aminopropyltriethoxysilane (10% in 25 mL dissolved water). The suspension was incubated at 80 ℃ for 2 h with constant mixing, washed thoroughly with water before drying at 120 ℃ for 2 h, and then 1%(v/v) glutaraldehyde was added to the silica gels suspended in 100 mM of phosphate buffer(pH 8) at 20 ℃. After stirring (150 rpm) for 2 h, the suspension was filtered and carriers were washed with water. The activated silica gels were resuspended in 100 mM of phosphate buffer (pH 8). Finally, a solution of enzyme was added. At this time, added amount of protein was almost equal to about 10 mg/mL. The suspension was stirred at 20 ℃ for 2 h and the immobilized GL-7-ACA acylase recovered by filtration was washed. After resuspending in 100 mM phosphate buffer (pH 8), the immobilized enzyme was analyzed.

Assay of GL-7-ACA acylase activity

The activity of the immobilized enzyme was measured as follows. The immobilized enzyme was incubated at 37 ℃ for 10 min in the presence of 1%(w/v) GL-7-ACA. The reaction was stopped with an aqueous solution of 20%(v/v) acetic acid and 0.05 N NaOH. Then ρ -dimethylaminobenaldehyde (PDAB, 0.5% w/v in methanol) was added to the mixture. The absorbance of mixture at 415 nm was then measured. One Unit of acylase activity was defined as the amount of enzyme that produced 1 mol of 7-ACA per min at 37 ℃, pH 8.

Determination of the degree of silanization

The degree of silanization was determined by ninhydrin method. The ninhydrin method was based on the method described by Park et al. (*17*). To

1mL of 3-APTES solutions in distilled water, 1 mL of different concentrations of ninhydrin reagent was added, and reacted in a covered boiling water bath at different heating times. The reactants were then cooled below 30°C in a cold-water bath and the contents were diluted with 5 mL of 50%(v/v) ethanol/water.

The absorbance level at 570 nm was measured with a UV/VIS spectrophotometer (Youngwoo Corp., Seoul, Korea).

Determination of the amount of protein bound to the carriers

The amount of protein was determined by the Folin-Lowry method (*18*). The amount of protein bound to the carriers was determined by the difference between initial and residual protein concentrations.

Calculation of yields

Yields of each step in the immobilization process were calculated as the ratio of the amount bound on the silica gels to the initial amount. Yields were expressed as a percentage.

Results and Discussion

In this study, GL-7-ACA acylase was immobilized on the silica gels by crosslinking with glutaraldehyde. The overall immobilization procedure introduced in this study has several steps such as pretreatment of silica gel, modification of silica gel surface by silanization, crosslinking by glutaraldehyde and coupling of GL-7-ACA acylase on the silica gel activated by glutaraldehyde. Thus, each step in the immobilization procedure was optimized individually to improve the efficiency of overall immobilization procedure of GL-7-ACA acylase. We also investigated the effect of sodium borohydride in order to increase the stability of the linkage between enzyme and support after immobilization.

Pretreatment of silica gel

Pretreatment of carrier serves to eliminate surface contaminants and to activate the surface silanol groups so that they are available for reaction with the silanating agent.

The literatures on pretreatment of carriers have been reported by many researchers and various chemicals has been used in the pretreatment of carrier(8,19). In this study, the effect of pretreatment was investigated with various chemicals reported in previous literatues in order to determine the suitable chemical for pretreatment of silica gels. As shown in Figure 1, both yields of silanization and protein binding obtained by silica gels pretreated with 35% of hydrogen peroxide were higher than those obtained by the other chemicals. Furthermore, the highest activity of immobilized GL-7-ACA acylase was also obtained from silica gel pretreated with 30% of hydrogen peroxide. Thus, we decided 30% of hydrogen peroxide as chemical for pretreatment.

Silanization on the silica gel surface

One of the most important techniques for enzyme immobilization is the attachment of amino groups through modification of the carrier surface (8). Silanization is the most common and effective method for the modification of the carrier surface. Organosilane agents containng amino groups are widely used in silanization because its amino groups are susceptible to the coupling reaction.

Three different agents, 3-aminopropyltriethoxysilane(3-APTES), 3-aminopropyltrimethoxysilane (3-APTMS) and 3-(trimethoxysilyl) propylethylen diamine (3-TMSPEDA), were tested in this study. As shown in Figure 2, the best results of the protein binding yield and activity of immobilized GL-7-ACA acylase were obtained by using 3-APTES. Thus, 3-APTES was chosen as a suitable reagent for silanization and used in further studies.

Aqueous organic solvents have usually been introduced in silanization to enhance the efficiency of silanization (20). The results in Table I indicated that silanization has no relation with polarity of sovent used. The highest yield of silanization was obtained when silanization was carried out in acetone, and protein binding yield and activity of immobilized GL-7-ACA acylase were also higher than those obtained by others.

However, reaction temperature of 80 °C used in silanization is unsuitable to immobilization procedure because boiling point of acetone is 56 °C. Such a reaction at above boiling point generates a serious problem such as increasing of pressure in immobilization process, and results in the difficulty of industrial applications. Thus, silanization was carried out at the temperature lower than the boiling point of acetone. As shown in Table I, fortunately, yield of

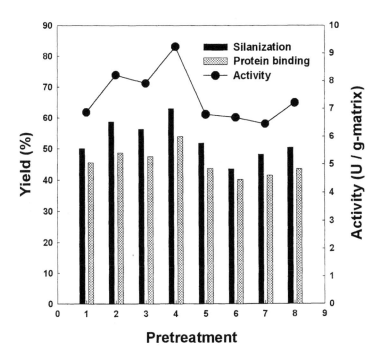

Figure 1. Effect of pretreatment of silica gel on the immobilization of GL-7-ACA acylase. Experiments were carried out with (1) distilled water, (2) 1.5N HNO₃ at 80°C for 2 hour, (3) Mixture of H₂SO₄ and H₂O₂ at 80°C for 2 hour, (4) 35% H₂O₂ at 25°C for 2 hour, (5) 30% H₂SO₄ at 80°C for 2 hour, (6) 10N NaOH at 80°C for 2 hour, (7) 5N NaOH at 80°C for 2 hour and (8) sonicating in distilled water, acetone and ethanol for 15min each.

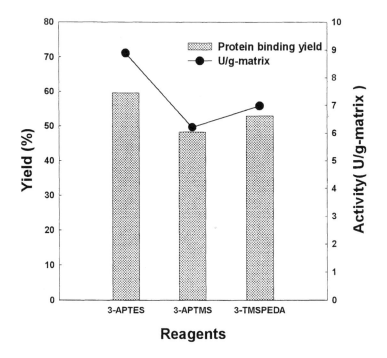

Figure 2. Effect of reagent on the silanization . Experiments were carried out with 3-aminopropyltriethoxysilane(3-APTES), 3-aminopropyltrimethoxysilane(3-APTMS) and 3-(trimethoxysilyl) propylethylendiamine (3-TMSPEDA).

silanization obtained at 50 °C is almost same as that obtained at 80 °C. Protein binding yield and activity of immobilized GL-7-ACA acylase obtained at 50 °C were also similar to that obtained at 80 °C. Thus, we decided 50 °C as a temperature of silanization using 3-APTES in acetone. Although not shown in this study, the reaction time of silanization and concentration of 3-APTES may also be important factors. It was determined as 120 min and 10%, respectively, in our previous study.

Table I. Effect of solvent on silanization

olvent	Polarity	Yield of silanization (%)	Yield of protein binding (%)	Activity of immobilized GL-7-ACA acylase (U/g-matrix)
Water	9.0	65.8	51.9	8.0
Dimethylsulfoxide	7.2	69.9	61.7	8.1
Acetone	5.1	80.8	72.9	11.9
Ethanol	5.2	70.8	60.9	7.9
Ethyl acetate	4.4	72.4	64.7	8.4
Iospropanol	3.9	73.6	64.9	9.0
Dichromethane	3.1	69.4	58.5	7.9
Xylene	2.5	68.9	53.8	7.7
Toluene	2.4	73.0	65.6	9.6
Hexane	0	51.8	40.8	5.9

Table II. Effect of temperature on silanization in acetone

Temperature (°C)	Yield of silanization (%)	Yield of protein binding(%)	Activity of immobilized GL-7-ACA acylase (U/g-matrix)
80 °C	82.6	75.58	12.1
50 °C	81.2	76.54	11.9
25 °C	47.2	53.0	5.16

Crosslinking by glutaraldehyde

Glutaraldehyde is a bifunctional reactive agent mainly capable of reacting with the surface amine groups of enzyme and carriers, through the formation of Schiff bases and Michael adducts (21). However, glutaraldehyde gives a negative effect such as toxicity to the enzyme and induction of enzyme denaturation to the immobilized enzyme. Therefore, glutaraldehyde was modified in aqueous solution of pH 8.0 at 62 °C for 25min in order to overcome the negative effect of glutaraldehyde. As shown in Figure 3, when using the non-modified glutaraldehyde, yield of protein binding increased with increasing glutaraldehyde. Nevertheless the activity of immobilized GL-7-ACA acylase gradually decreased at above 1% (v/v) of glutaraldehyde. On the other hand, when using the modified glutaraldehyde, the behavior of protein binding yield is very similar to that of non-modified glutaraldehyde. However, unlike non-modified glutaraldehyde , the activity of immobilized GL-7-ACA acylase increased up to over 2% concentraion of modified glutaraldehyde, then almost unchanged afterwards. Activity of immobilized GL-7-ACA acylase at optimal condition was 18.72 Unit/g-matrix. This value was increased up to about 63 % compared to that obtained by using non-modified glutaraldehyde.

Based on above results it may be stated that crosslinking by modified glutaraldehyde resulted in enhanced stability to compared non-modified glutaraldehyde. Namely, these results imply that the negative effect of glutaraldehyde was reduced by modification of glutaraldehyde.

Effect of enzyme concentration on immobilization

To optimize the amount of enzyme added, it is important to determine the capacity of the carrier for the retention of enzyme activity. The effect of GL-7-ACA acylase concentraion added during the immobilization step is shown in Figure 4. The amount of GL-7-ACA acylase immobilized on 100 mg of the silica gels increased with increasing concentration of GL-7-ACA acylase, and reached a plateau at a concentration of 10 mg/ml. Also, the behavior of activity of immobilized GL-7-ACA acylase is almost same as amount of GL-7-ACA acylase immobilized on silica gels. Therefore, 10 mg/ml was chosen as an optimal concentration of GL-7-ACA acylase per 100 mg of the silica gels added in immobilization procedure.

Improvement of stability of immobilized GL-7-ACA acylase by treatment of reducing agent

In order to increase the stability of the linkage between enzyme and

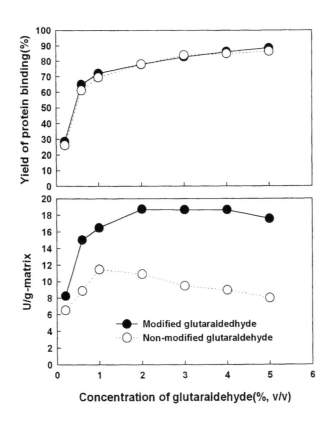

Figure 3. Effect of glutaraldehyde on the immobilization of GL-7-ACA acylase.

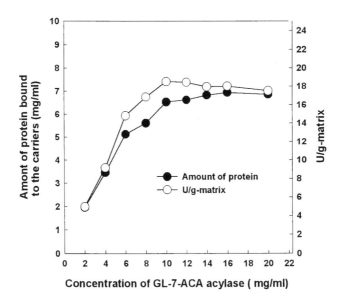

Figure 4. Effect of GL-7-ACA acylase concentraion on the immobilization (Silica gels : 100mg).

support, we investigated the effect of reducing the Schiff's bases formed in the glutaraldehyde coupling (22). Sodium borohydride was selected as reducing agent. As shown in Figure 5, The treatment with sodium borohydride did not significantly affect the activity of the immobilized enzymes until the concentrations of sodium borohydride increased up to 2 % (w/w-carrier), determining a loss of activity < 5 %. Beyond this concentration, activity of immobilized enzyme decreased with increasing concentration of sodium borohydride.

On the basis of the above results, the abilities for long-term stability of the immobilized GL-7-ACA acylase treated with 2% (w/w) sodium borohydride were investigated by repeated washes. As shown in Figure 6, the effect of sodium borohydride reduction on operational stability was obvious. The activity of immobilized GL-7-ACA acylase treated with 2% (w/w) sodium borohydride was remained almost 86 % after 20 times of reuse.

As a consequence, if this immobilization method should be introduced in scale-up for the mass production of 7-ACA, more superior effects might be obtained.

Conclusions

Immobilization of GL-7-ACA acylase to silica gel via silanization followed by glutaradehyde was introduced in this study. Silica gels were pretreated with various chemicals. Hydrogen peroxide (35%) among various chemicals tested was found to be the best one for the immobilization of GL-7-ACA acylase.

The silanization of the silica gel surface is a well-known and convenient method for combining silica gel with an organic molecule. Three different reagents were tested in this study. 3-APTES among the reagents tested was chosen as a suitable reagent for silanization. On the other hand, effect of solvent on silanization was also investigated, and the best result was obtained when the silanization was carried out in acetone at 50 °C.

Glutaraldehyde has been widely used as a crosslinking agent for the immobilization of enzymes. However, difficulty lies in obtaining reproducible activity of the immobilized enzyme because certain adverse effects of glutaraldehyde have existed. Glutaraldehyde was modified in aqueous solution of pH 8.0 at 62 °C for 25min in order to overcome the negative effect of glutaraldehyde. Activity of the immobilized GL-7-ACA acylase when using the

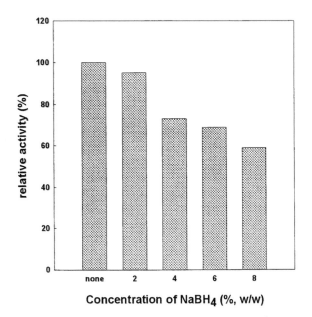

Figure5. Effect of NaBH₄ on the immobilization of GL-7-ACA acylase. Activities are expressed as percentage to the immobilized GL-7-ACA acylase untreated with NaBH₄. 19.56 Unit/g-matrix corresponds to the relative activity of 100%.

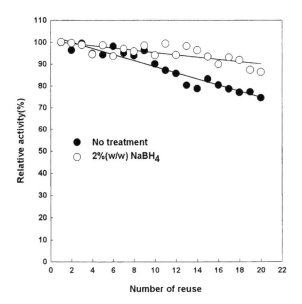

Figure 6. Operational stability of GL-7-ACA acylase immobilize on silica gels. 19.12 Unit/g-matrix of immobilized GL-7-ACA acylase untreated with NaBH₄ corresponds to the relative activity of 100%, and 18.81 Unit/g-matrix of immobilized GL-7-ACA acylase treated with 2% NaBH₄ corresponsd to the relative activity of 100%.

modified glutaraldehyde was 18.72 Unit/g-matrix. This value was increased up to about 63 % compared to that obtained by using non-modified glutaraldehyde. To optimize the amount of enzyme added, it is important to determine the capacity of the carrier for the retention of enzyme activity. In this study, 10 mg/ml was determined as optimal concentraion of GL-7-ACA acylase per 100 mg of the silica gel.

We also investigated the effect of sodium borohydride in order to increase the stability of the linkage between enzyme and support after immobilization. The activity of immobilized GL-7-ACA acylase treated with 2% (w/w) sodium borohydride was remained almost 86% after 20 times of recycles.

Acknowledgements

This study was partially supported by research grants from the KOSEF through the Applied Rheology Center at Korea University, Korea.

References

1. Alfani, F.; Cantarella, M.; Cutarella, N.; Gallifuoco, A.; Golini,P.;Bianchi, D. *Biotech .Lett*. **1997,** *19,* 175-178.
2. Bianchi, D.; Golini, P.; Bortolo, R.; Battistel, E.; Tassinari, R.; Cesti, P. *Enzyme Microb. Technol.* **1997,** *20,* 368-372.
3. Ezio, B.; Daniele, B.; Rossella, B.; Lucia, B. *Appl. Biochem. Biotech.* **1998,** *69,* 53-67.
4. Cowan, D.A.; Daniel, R.M. *Biotechnol. Bioeng.* **1982,** *14 ,* 2053-2061
5. Antonio, R.C.; Claudio, A.; *Langmuir,* **1997,** *13 ,* 2681-2686.
6. Cestari,A.R.; Vieira, E.F.S.; Simoni, J.A.; Airoldi, C. *Thermochim. Acta,* **2000,** *348,* 25-31.
7. Vieira, F.S.; Cestari, A.R.; Simoni, J.A.; Airoldi, C. *Thermochim. Acta,* **1999,** *328,* 247-252.
8.Flounders, A.W.; Brandon, D. L.; Bates, A.H.; *Appl. Biochem. Biotechnol.* **1995,** *50,* 265-283.
9. Nanci, A.; Wuest, J.D.; Peru, L.; Brunet, P.; Sharma, V.; Zalzal, S.; Mckee, M.D. *J. Biomed. Mater. Res.* **1998,** *40,* 324-335.
10. Matsuda,S.; Iwata,H.; Se, N.; Ikada, y. *J.Biomed. Mater.Res.* **1999,** *45,*20-27.
11. Wilhem, T.; Frank, W. *Top. Curr. Chem.* **1999,** *200,* 95-126.

12. Terumichi, N.; Kenji, I.; Shigemasa, Y.; Akimasa, S.; Hisashi, T.; Toshio, T.; Masahiro, I. *Chem. Pharm. Bull.* **1989,** *37,* 2463-2466.
13. Alfani, F.; Cantarella, M.; Cutarella, N.; Gallifuoco, A.; Golini, P.; Bianchi, D. *Biotech .Lett*. **1997,** *19,* 175-178.
14. Bianchi, D.; Golini, P.; Bortolo, R.; Battistel, E.; Tassinari, R.; Cesti, P. *Enzyme Microb. Technol.* **1997,** *20,* 368-372.
15. Hublik, G.; Schinner, F. *Enzyme Microb. Technol.* **(2000),** *273,* 30-336.
16. Shibuya ,Y.; Matsumoto, K.; Fujh, T. *Agric. Biol. Chem.* **1981,** *45,* 2225-2229.
17. Park, S.W.; Kim, Y.I.; Chung, K.H.; Kim, S.W. *J. Micorbiol. Biotechnol.* **2001,** 11, 199-203.
18. Lowry, O. H.; Rosebrough, N.L.; Farr, A.L.; Randall, R.J., *J. Biol.Chem.* **1951,** *193,*265-275.
19. Aravind, S.; Stephen, J. K.; Patrick, I.O.; Jacobson, K.B.; Jonathan, W.; Michel, J.D. *Enzyme Microb.Technol.* **1999,** *24,*26-34.
20. Bobinson, P.J.; Dunnil, P.; Lilly, M.D.; Biochim. Biophys. Acta. **1971,** *242,* 659-665.
21. Monsan, P. J. *Mol. Catal.* **(1997/78),** *3,* 371-384.
22. Bianichi, D.; Golini, P.; Bortolo, R.; Cesti, P. *Enzyme Microb. Technol.* **1996,** *18,* 592-596.

Chapter 9

Production of Laccase by Membrane-Surface Liquid Culture with Nonwoven Fabric of *Coriolus versicolor*

Kazuhiro Hoshino[1], Mami Yuzuriha[2], Shoichi Morohashi[1], Shigehiro Kagaya[1], and Masayuki Taniguchi[3]

[1]Department of Material Systems Engineering and Life Science, Toyama University, Toyama 930–8555, Japan
[2]Department of Chemical and Biochemical Engineering, Toyama University, Toyama, Japan
[3]Department of Materials Science and Technology, Niigata University, Niigata 950–2181, Japan

The membrane-surface liquid culture with nonwoven fabric (MSLC-NF) was developed in order to produce effectively laccase from *Coriolus versicolor* IFO4937 and to apply to the bioremediation of environmental pollutants. In the batchwise production of laccase using MSLC-NF as the naringin was added as an inducer, the cell was well growth on the nonwoven fabric and the amount production of laccase was as the same as that by the normal surface culture. In the continuous production of laccase using MSLC-NF by supplying the fresh medium at a dilution rate of 0.05-0.10 h^{-1}, laccase could be produced successively for 50 days. Further, when this system was applied to the continuous bioremediation of the wastewater contained 0.1-1.0 mM of 2,4,6-trichlorophenol, the pollutant in the wastewater was almost completely degraded for 50 days.

© 2002 American Chemical Society

109

Laccase (oxidoreductase, EC 1.10.3.2) have been found in fungal stains belonging to various classes (1-5). Recently, lignin biodegradation by laccase has attracted the interest of many researchers, who aim to develop applications of lignolytic fungi and enzyme systems for pretreatment of the lignocellulic materials (6-8). Furthermore, numerous reports show that laccase secreted by white-rot fungi have been received attention because chlorined aromatic compounds used as biocides or wood preservatives can be dechlorinated and detoxified (9-11). However, the apparatus and the cultivation method for the production of laccase and the degradation of highly toxic environmental pollutants using the fungi have been hardly examined. The fungi grow well under lower moisture conditions than those required for optimal bacterial growth. Further, they require an aerobic environment for the production of specific products and grow well on the surface of wood found in the natural environment (12,13). Thus, many reports of fungi production of useful substances in much larger amounts by solid-state culture than by shaking cultures have been published (12-15). However, the solid-state cultures have several weak points, such mass transfer limitation of substrate and difficulty of control of culture conditions. To overcome these problems, Ogawa et al. proposed a membrane-surface liquid culture (MSLC) (16,17). The apparatus which fungi can be grown on a microporus membrane surface exposed to the air, with the other side of the membrane in contact with liquid medium is used. However, in the bioremediation of wastewater using the lignolytic fungi the utilization of MSLC is quite difficult because the membrane used for the method is very expensive and most of enzymes secreted by the fungi were drained from the bioreactor. Therefore, to achieve the efficient degradation of environmental pollutants we conceived that the nonwoven fabric, which are cheaper than a microporus membrane such as ultrafiltration membrane and microfiltration membrane and are able to adsorb the secretases, was used for the MSLC. In this study, we developed a MSLC using nonwoven fabric (MSLC-NF) instead of microporus membrane. We applied the MSLC-NF to the continuous laccase production using *Coriolus versicolor*. In addition, the continuous biodegradation of 2,4,6-trichlorophenol in wastewater using MSLC-NF was examined.

Materials and Methods

Microorganism and media   The white-rot basidiomycete fungi *Corilous versicolor* IFO4937 was used to produce laccase in this study. The vegetative cells were cultivated for 5 day of potato-dextrose agar at 28℃ to harvest mycelium. The grown mycelium was shaved aseptically from the agar and suspended in 0.85% saline. The basic medium for production of laccase contained 10 g glucose, 10 g polypepton (Nihon Seiyaku Co., Tokyo), 1.5 g KH_2PO_4, 0.5 g $MgSO_4 \cdot 7H_2O$, 16 mg $CuSO_4 \cdot 5H_2O$, and 2.0 mg Thiamin·HCl in 1 l of tapped water and the pH was adjusted to 5.6 with 1N HCl. The medium was autoclaved at 121℃ for 15 min.

Batchwise incubation conditions   Batchwise incubation experiments were carried out in stationary 100-ml Erenmyer flasks contained 25 ml medium, to which a nonwoven fabric (MB-T3, Japan Villene Co., Tokyo) were installed in flask to support and immobilize mycelia of *C. versicolor*. MB-T3 (T3) is mode of polyester having a thickness of 8 mm and an average mesh of 270 μm. The mycelium suspension was inoculated uniformly onto the nonwoven fabric. The incubation temperature was carried out at 28℃ for 20-21 days.

Membrane-surface liquid culture with nonwoven fabric (MSLC-NF)
Figure 1 shows a schematic diagram of the cultivation system for the membrane-surface liquid culture with nonwoven fabric (MSLC-NF). The cultivation apparatus was adopted the membrane-liquid surface culture (MSLC) proposed by Ogawa *et al.* (16,17) by using a nonwoven fabric instead of a microporus membrane.   The apparatus was made of polycarbonate and composed of a cylinder equipped with inlet and outlet (10 cm in diameter and 4.0 cm in height) and a cap holed for aeration.   About 100 ml of the liquid medium was poured into the apparatus and the nonwoven fabric cut in a circle was floated in the medium. After the cup was putted on, the apparatus was autoclaved at 121℃ for 15 min.   Continuous production of laccase was carried out as follows.   Four microliters of a suspension of mycelium was inoculated uniformly onto the nonwoven fabric, and incubated statically.   After the mycelium was covered perfectly on the nonwoven fabric for 5-6 days, the continuous culture was performed by supplying fresh medium at a certain dilution rate (*D*) more than 42 days. The medium supplied was twentieth part of the basic medium as a model wastewater.   In the continious bioremediation of wastewater, 2,4,6-trichlorophenol (2,4,6-TCP) was added into the medium as an environmental pollutant.

Enzyme assay   Enzyme activities were determined spectrophotometerically at pH 6.0 and 30 ℃ .   Laccase activity was assayed by the oxidation of syringaldazine which caused an increase in absorbance at 527 nm (18). One unit of enzyme activity was defined as 1 mol of syringaldazine oxidation per second.

Analysis   The dry cell weight was measured as follows.   In the shaking culture and surface liquid culture without nonwoven fabric, the dry cell weight was determined by subtracting the dry weight of filter paper from total dry weight of cells and the filter paper.   In the cultures using nonwoven fabric, the cells formed on the nonwoven fabric could be removed quite easily with a spatula.   The cells were weighed without additional washing.   The recovered cells were dried at 105℃ to a constant weight.   The glucose concentration was measured by the mutarotase GOD method (Glucose CII-Test; Wako Pure Chemicals, Osaka).   The environmental pollutants in medium was analyzed by

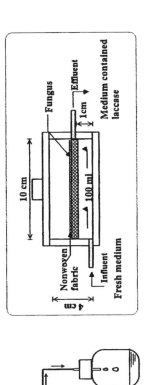

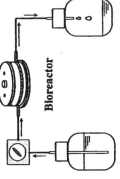

Figure 1 Schematic diagram of the cultivation system for MSLC-NF

HPLC using a Wakosil Agri-9 column (4.6 mm i.d. x 250 mm; Wako) using water/methanol (3:7, v/v) as eluate and monitoring at 290 nm and 35℃.

Result and discussion

Laccase production by *C.versicolor*

Figure 2 show the time course of cell growth, glucose consumption, and the production of laccase in shaking culture and surface culture of *C.versicolor*. In the both cultures, the dry cell weight increased as the glucose decreased and reached a maximum of 6.2 g-dry cells/*l* at incubation time of 12 day. On the other hand, the activities of laccase appeared after 8-day inculcations, increased with time and reached maximum values of 6 nkat/ml in shaking culture and 24 nkat/ml in surface culture at 20 day. No signification differences in the cell growth or glucose consumption were observed between the shaking culture and surface culture, though the activities in the later culture were significantly higher than those in the former. The reason seems to be that in a shaking culture the synthesis of laccase by the cell is limited due to a lack of oxygen resulting from cell sedimentation and a breakup of mycelium by shearing force resulting from shaking, whereas in a surface culture the cells can synthesize large amounts of laccase because the upper part of mycelium was exposed to air and undamaged by shearing force. Therefore, the surface culture was more effective than the shaking culture for the production of laccase by *C.versicolor*.

Induction of laccase by naringin

Many aromatic compounds have been widely used to stimulate production of laccase. 2,5-Xylidine was reported to be the most effective inducer for the production of laccase with *Polyporus versicolor* (19, 20). Laccase production from *Trametes versicolor*, *Pleurotus ostreatus* and *Pholiota mutabilis* is also increased by ferulic acid (21). *p*-Anisidine is other examples of inducers that enhance production of laccase from *Rhizoctonia praticola* (9). However, most of aromatic compounds are either very harmful to humans or at least expensive, precluding their use from industrial application. Therefore, we have been researched new inducer for effective laccase production by *C.versicolor*. As a result, we found out that naringin which was a kind of the flavonoidis is an effective inducer for laccase secretion. Figure 3 show the production of laccase by surface culture of *C.versicolor* when naringin was added into the medium at a certain concentration in the initial cultivation. The cell growth was about the same in spite of conditions. However, in the laccase production, the larger the amount of naringin added is, the higher the activity laccase is. In 3.5 mM naringin, the maximum activity was about 420 nkat/ml at 12-16 cultivation days, which was over 16 times higher than that without naringin. The addition of naringin into the medium is a high potential method to induce laccase because of low toxicity and popular price. Moreover, the method could be applied to the degradation of environmental pollutant, to which the reduction of the cost is indispensable.

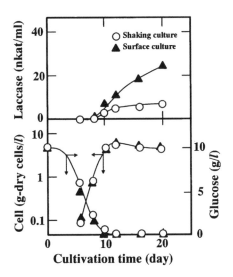

Figure 2   Production of laccase from C.versicolor by shaking culture and surface culture

114

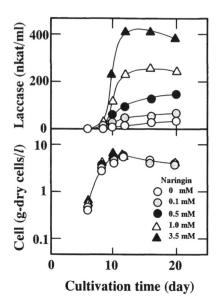

Cultivation time (day)

Figure 3 Induction of laccase by adding naringin

Batchwise production of laccase using surface culture and MSLC-NF

    Many report of fungi production of useful substances in mach larger amounts by solid-state culture than by shaking cultures have been published (12-15). However, the solid-state culture has several disadvantages, as mass transfer limitation inside the solid substrate and difficulty of the control of substrate concentration as well as pH of the medium during cultivation. To overcome these disadvantages of the solid-state culture, Ogawa *et. al.* proposed a new cultivation method, membrane-surface liquid culture (MSLC). They showed that the amount of neutral protease and kojic acid of *Asperigillus oryzae* using MSLC (16, 17) was much higher than those by shaking culture and the control such as pH, concentrations was easy. Although the MSLC have a high potential for the production of useful substances using filamentous fungus, it is undesirable to apply to the biodegradation of environmental pollutant as described above. Thus, to employ MSLC as a culture method for biodegradation of environmental pollutant, we proposed MSLC with a nonwoven fabric instead of a microporus membrane, and named membrane-surface liquid culture with nonwoven fabric (MSLC-NF).

    First, the laccase production by MSLC-NF was compared to that by surface culture. Figure 4A and 4B show the result of surface culture and MSLC-NF, respectively. Both cultures were carried out in the basic medium contained 1mM naringin. In the both culture, the dry cell weight increased as the glucose

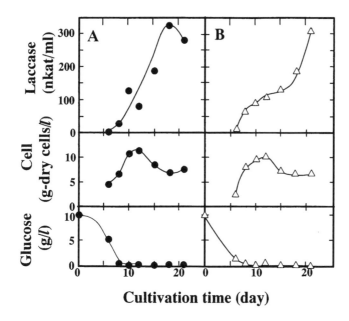

Cultivation time (day)

Figure 4 Production of laccase by surface culture and MSLC-NF

consumed and reached a maximum (12.4 g-dry cells/l in SC, 10.9 g-dry cells/l in MSLC-NF) at an incubation time of 12 day. In the MSLC-NF, the cells was grown like a sheet on the nonwoven fabric and were not observed in liquid medium under the fabric at all. In the surface culture, the activities of laccase were appeared after 6-day incubation, reached maximum value of 320 nkat/ml at an incubation time of 18 day in MSLC-NF. However, the production rate of laccase in MSLC-NF was later than that in surface culture. The maximum activity was 300 nkat/ml at an incubation time of 21 day. It is thought that the late induction of activity in MSLC-NF is attributed to the limitation of the mass transfer to mycelium grown on the nonwoven fabric. However, the enzyme secreted from the fungi was adsorbed on the fabric recovered after 21 day of cultivation in the MSLC-NF (0.8 mg-protein/g-dry fabric) and the fabric have a high laccase activity (34 nkat/g-dry fabric). This characteristic is not reported in the papers described the application of MSLC using a microporus membrane (16,17), and it is suggested that the nonwoven fabric can be used as carrier for immobilization of enzyme in addition to the production of useful substance.

Continuous production of laccase by MSLC-NF
    To enhance the laccase productivity, we tried to produce laccase continuously by MSLC-NF shown in Figure 1. Since excess supply of glucose

116

resulted in the decline of its production in preliminary test, the supply medium diluted the basic media to 1/20. Figure 5A shows the results of the continuous production of laccase using MSLC-NF without naringin. After the cell was covered on the nonwoven fabric at an incubation time of 5 day, the continuous cultivation was started at a dilution rate of 0.1 h^{-1}. The laccase in the effluent showed little activity (average 0.05 nkat/ml) for 12 days. Thus, when the dilution rate was changed at 0.05 h^{-1} on 12 day, the activity increased rapidly and 0.4-1.1 nkat/ml laccase was constantly produced for 38 days in spite of a little adsorption of secretases on the fabric. In this phase of laccase production, the laccase productivity was about 26.5 nkat/l/day. On the other hand, as the medium contained 1 mM naringin was supplied as shown in Fig.5B, the activities was more than twice time as high as those without naringin. In the operation condition of dilution rate 0.05 h^{-1}, the productivity of laccase produced was average 60 nkat/l/day, which is 2.2-fold higher than that of continuous MSLC-NF without naringin. During all continuous MSLC-NFs, little or no glucose was detected except for the initial period and the the living cells cannot be observed in the effluent at all. Accordingly, the utilization of MSLC-NF allowed to be living the mycelium on the nonwoven fabric for a long time and to product consecutively laccase. The MSLC-NF is expected to be a promise method for the continuous production of useful substance such as organic acids, enzymes, antibiotics, *etc.* using filamentous fungus as well as the continious MSLC reported by wakisaka *et al.* (22).

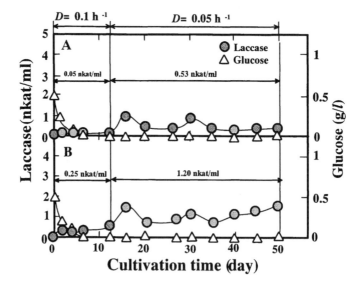

Figure 5 Continuous production of laccase by MSLC-NF

Continuous degradation of 2,4,6-trichlorophenol in wastewater using MSLC-NF

Since it is recently clarified that oxidoreductase such as laccase, lignin peroxidase, Mn peroxidase *etc.* are capable of degrading highly toxic environmental pollutants, the technique of bioremediation using these enzymes has been investigated (9,10, 23-27). However, those technologies are not so suitable for continuous processing of wastewater because the immobilization of laccase on water insoluble carrier is necessary. Moreover, if the immobilized laccase is an inactive, the preparation should be exchanged on the way to the treatment. To overcome these problems, we examined the application of MSLC-NF to the continuous treatment of wastewater contained 2,4,6-TCP. The diluted medium contained 0.1-1.0 mM 2,4,6-TCP as a model wastewater was supplied at a diluted rate of 0.1 h^{-1}. The other procedures for continuous operation were similar to that for continuous production of laccase. Figure 6 shows the results of continuous treatment of wastewater contained 2,4,6-TCP. In order to clarify the adsorption of 2,4,6-TCP on nonwoven fabric, the case where the cells is not

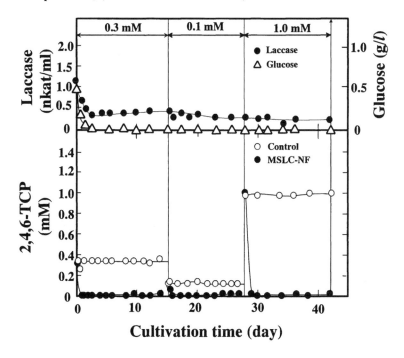

Cultivation time (day)

Figure 6 Continuous degradation of 2,4,6-TCP using MSLC-NF

118

inoculated on the fabric was also examined (Control in Figure 6). It is thought that there is little adsorption of 2,4,6-TCP on the fabric because in the control the concentration in effluent was almost equal to that in influent during the continuous operation. On the other hand, after the continuous opration was initiated by supplying the wastewater contained 0.3 mM 2,4,6-TCP, the concentration in the effluent decreased rapidly and was at least 0.01 mM or less until 16 day. This tendency was the same though changed the concentration of 2,4,6-TCP in the wastewater through the cultivation time of 50 days. Further, no or little glucose in the effluent was detected during this culture but laccase in the effluent was continuously produced at the level of 0.3-0.4 nkat/ml for 50 days, which is about one-forth that in the continuous laccase production by MSLC-NF. However, the results indicate that the cell that was exposed to pesticide for a long time is surviving. From the result in which the nonwoven fabric recovered after the cultivation was finished have a high laccase activity, it was suggested that laccase immobilized on the fabric as well as free laccase was responsible for a part of biodegradation of 2,4,6-TCP. Accordingly, it is verified that the continuous MSLC-NF could be used for bioremediation of wastewater contained environmental pollutants as well as for production of useful substances.

Conclusion
    The fungi are used not only for production of traditional fermentative foods but also for the production of enzymes, antibiotics, beverages, and volatile compounds. Further, the biodegradation of toxic environmental pollutants by oxidoreductase secreted the white-rot fungi have been recently received attention. However, the cultivation method to give the physiological properties of the fungi full play has been hardly investigated. Therefore, we proposed the membrane-surface culture with nonwoven fabric (MSLC-NF), in order to produce laccase from white-rot fungus and to apply to the bioremediation of environmental pollutants. When the continuous laccase production was performed using MSLC-NF, cell was well growth on the nonwoven fabric and laccase could be produced successively for 50 days. Moreover, when this system was applied to the continuous bioremediation of the wastewater contained 2,4,6-TCP, the pollutant in the effluent was almost completely degraded for 50 days. The continuous system used the MSLC-NF suggests a potential not only for the production of secreting enzyme from fungi but also for the bioremediation of the wastewater contained various pollutants.

Acknowledgments

    The authors wish to thank for Japan Villene Co. supplying nonwoven fabrics T3. This study was supported by Grant-in-Aid for Scientist Research

(No.13027228) from the Ministry of Education, Science, Sports, and Culture (Japan).

References

1. Bollang, J.; Leonowicz, A. *Appl. Environ. Microbiol.* **1984**, *48*, 849-854.
2. Sandhu, D. K.; Arora, D. S. *Experientia, 1985*, 41, 355-356.
3. Oda, Y.; Adachi, K.; Aita, I.; Ito, M.; Aso,Y.; Igarashi, H. *Agric.Biol.Chem.* **1991**, *55*, 1391-1095.
4. Nishizawa, Y.; Nakabayashi, K.; Shinagawa, E. *J. Ferment. Bioeng.* **1995**, *80*, 91-93.
5. Coll, P. M.; Fernández-Abalos, J. M.; Villanueva, J. R.; Stntamariá, R.; Pérez, P. *Appl. Environ. Microbiol.* **1993**, *59*, 2607-2613.
6. Gold, M. H.; Alic, M. *Microbiol. Rev.* **1993**, *57*, 605-622.
7. Fetzner, S.; Lingens, F. *Microbiol. Rev.* **1994**, *58*, 641-685.
8. Eggert, C.; Temp,U.; Dean, J. F. D.; Eriksson, K.-E. L. *FEBS Lett.* **1996**, *391*, 144-148.
9. Bollag, J.-M.; Shuttleworth, K. L.; Anderson, D. H. *Appl. Environ. Microbiol.* **1988**, *54*, 3086-3091.
10. Ruggiero, R.; Sarkar, J. M.; Bollang, J.-M. *Solid Sci.* **1989**, *147*, 361-370.
11. Lin, J.-E.; Wang, H. Y.; Hickey, R. F. *Biotechnol. Bioeng.* **1990**, 1125-1134.
12. Bajracharya, R.; Mudgett, R. E. Biotechnol. Bioeng., 1980, *22*,2219-2235.
13. Johns,M. R. In *Solid sbstrate cultivation;* Doelle, H. W., Mitchell, D.A., Rolz, C.E. Eds.; Elsevier Applied Science: London and New York, 1992; 341-352.
14. Grajeck, W. *Appl. Microbiol. Biotechnol.* **1987**, *26*, 126-129.
15. Sudo, S.; Ishikawa, T., Sato, K.; Oba, T. *J. Ferment., Bioeng.* **1994**, *77*, 483-489.
16. Ogawa, A.; Yasuhara, A.; Tanaka, T.; Sakiyama, T.,;Nakanishi, K. *J. Ferment. Bioeng.* **1995**, *80*, 35-40.
17. Ogawa, A.; Wakisaka, Y.; Tanaka, T.; Sakiyama, T.,;Nakanishi, K.; *J. Ferment. Bioeng.* **1995**, *80*, 41-45.
18. Leonowicz, A.; Grzywnowicz, K. *Enzyme Microb. Technol.* **1981**, *3*, 55-58.
19. Fåhraeus, G.; Tullander, V.; Ljunggren, H. *Physiol. Plant.* **1958**, *11*,631-643.
20. Fåhraeus, G.; Reinhammer, B. Acta Chem. Scand. 1967, 21, 2367-2378.
21. Leonowicz, A.; Trojanowski, J.; Orlicz, B. Acta Biochem. Polo. **1978**, ·25, 369-378.
22. Wakisaka, Y.; Segawa, T.; Imamura, K.; Sakiyama, T.; Nakanishi, K.; *J. Ferment. Bioeng.*, **1998**, *85*, 488-494.
23. Thomas, D.R.; Carswell, K. S.; Georgiou, G. *Biotechnol. Bioeng.*, **1992**, *40*, 1395-1402.

24. Bezalel, L.; Hadar,Y.; Cerniglia, C. *Appl. Environ. Microbiol.* **1996**, *62*, 292-295.
25. Valli, K., Gold, M. H. *J. Bacteriol.* **1991**, *173*, 345-352.
26. Kiyohara, H.; Hatta,T.; Ogawa, Y.; Kakuda, T.; Yokoyama, H.; Takizawa, N. *Appl. Environ. Microbiol.* **1992**, *58*, 1276-1283.
27. Lamar, R. T.; Larson, M. J.; Kentkiek, T. *Appl. Environ. Microbiol.* **1990**, *56*, 3519-3526.

Chapter 10

Hydrolysis of Paper Sludge Using Mixed Cellulase System: Enzymtic Hydrolysis of Paper Sludge

Sang-Mok Lee[1], Jianqiang Lin[2], and Yoon-Mo Koo[1,*]

[1]Department of Biological Engineering, Inha University, Incheon 402–751, Korea
[2]State Key Laboratory of Microbial Technology, Shandong University, Jinan 250100, China
*Corresponding author: Phone: 82–32–860–7513; fax: 82–32–875–0827; email: ymkoo@inha.ac.kr

The effects of water soluble materials in paper sludge (sludge for short) on cellulase and β-glucosidase activities and the effects of pre-drying of sludge and initial glucose concentrations on the enzymatic sludge hydrolysis were studied. The optimization of enzyme system for hydrolysis of the sludge for production of glucose was also made. Water soluble materials in the sludge showed stimulatory effect on carboxymethyl cellulose (CMC) activity, inhibitory effect on filter paper (FP) activity and no effect on avicelase and β-glucosidase activities. Re-wetted sludge after drying showed 64% decrease in glucose production than that for the non-dried sludge in hydrolysis. The initial glucose concentrations of 5, 10 and 20 g/l resulted in 9.5%, 20.4% and 39.2% decreases in glucose production, respectively, after 16 h of hydrolysis. CMC and β-glucosidase activities at 5 and 10, 10 and 10, 10 and 20 U/ml were optimal for hydrolysis of 5, 10 and 20% of sludge, respectively.

© 2002 American Chemical Society

Paper sludge has caused severe environmental problem. Utilizing the sludge as a production material in industry not only solves the environmental problem but also reduces the production cost in industry. The dried materials, comprising 40% of this sludge, are composed of 30 to 60% of cellulose, 5 to 10% of lignin, with the remaining composed of mainly ashes. Cellulose in sludge can be hydrolyzed chemically or enzymatically to produce glucose which in turn can be used as a substrate of fermentation for the production of fuel (1,2), organic acids (3) and other useful products. Chemical hydrolysis of cellulose yields waste acid or alkaline, which will eventually lead to environmental pollution, while enzymatic hydrolysis can be free from this drawback. Many works have been done on enzymatic hydrolysis of biomass, such as different types of wood (4) and rice straws (5). These natural materials need to be pretreated in order to become easily attacked by enzymes (4,5,6). The commercial development was hindered because the pretreatment process requires a lot of energy (7,8) and the enzyme is of higher cost than acid or alkaline. On the other hand, sludge might need no pretreatment because it has already been treated for removing lignin and hemicellulose and is much more susceptible to enzymatic hydrolysis (9). In this research, the enzyme system was optimized to obtain high glucose yield with relatively less enzymes to reduce the cost for the enzymes.

Cellulose is degraded by the synergistic action of three types of enzymes in the cellulase complex, namely, endoglucanase, exoglucanase and β-glucosidase. Endoglucanase acts randomly on the internal bond of amorphous cellulose to break the polymer. Exoglucanase cuts the cellulose polymers from their nonreducing terminals producing cellobiose unit. Exoglucanase is intensively inhibited by cellobiose. β-glycosidase degrades cellobiose into glucose that relieves the feedback inhibition of cellobiose on exoglucanase. Glucose, the final product, inhibits various steps breaking cellulose molecule into glucose (10). Therefore, it is necessary to employ a balanced enzyme system containing appropriate amounts of endo- and exo-type glucanase as well as β-glucosidase for achieving an efficient hydrolysis of cellulose. In this research, we investigated optimal conditions, especially the balance of the enzyme system for the hydrolysis of paper mill sludge.

MATERIALS AND METHODS

Enzymes and Reaction Conditions

Raw cellulase powder (11) produced by *Trichoderma reesei* Rut C-30 (ATCC 56765) and commercial β-glucosidase solution (Novozym 188, Novo Nordisk, Denmark) mixed each other in various compositions were used in this research. Certain amount of sludge and enzymes in 20 ml citrate buffer (pH 4.8) in 100 ml flasks were incubated at 50°C and 350 rpm in a shaking water bath

(KMC1205KW1, KMC Vision Co., Korea). Oxymycin, an antibiotic, was added to make final concentration of 25 mg/ml to prevent contamination. The flasks containing the mixture of sludge and buffer solution were sterilized at 121 °C for 30 min and then cooled down to about 50 °C before adding the enzymes and antibiotics. Sludge was obtained from Samwha Paper Co., Korea.

Analytical Methods

CMC and FP degrading activities, avicelase and β-glucosidase activities were determined following the method of the International Union of Pure and Applied Chemistry (IUPAC). One unit of the enzyme activity was defined as the amount of enzyme releasing 1 μmol of glucose per minute. Glucose concentration in enzymatic hydrolysis experiments was measured using the glucose analyzer (YSI 2700, Yellow Springs Instrument Co., USA).

RESULTS AND DISCUSSION

Effect of Water Soluble Materials in Sludge on Enzyme Activities

A sludge of 5% (dry weight/volume) was washed using distilled water by stirring at about 500 rpm for 2 h, and then centrifuged at 4 °C, 8000 rpm for 30 min to recover the sludge washing water. This water was then used to dissolve the cellulase powder and the β-glucosidase solution (described in Materials and Methods) for activity measurement. The controls of cellulase and β-glucosidase dissolved in distilled water were also made for enzyme activity measurement. The results of cellulase and β-glucosidase activity measurement were shown in Table 1. It shows that water extractable materials in the sludge has stimulating effect on CMCase activities but inhibitory effect on FPase activities while negligible effect was shown on avicelase activities. Water extractable materials also have no effect on β-glucosidase activities.

Paper-mill industries use numerous chemicals during the manufacturing processes including recycling papers which might have toxic effects on enzymes. Paper cellulose are purified by bleaching with chlorine, alkaline and sometimes hypochlorite. Other additives are used in the preparation of pulp, such as kaoline, talc, titanium dioxide, calcium carbonate, zinc oxide and sulfate, diatomaceous earth etc., whereas gelatine, vegetable glue, etc. are used as binders. About 50% of the dry sludge is constituted by organic matter. The mineral component is mainly calcium carbonate, talc and kaolin, and the metals are mostly aluminum and zinc. Although many chemicals, which might be toxic to enzymes, remained in the sludge, the above result showed that these chemicals do not affect the enzyme activities significantly. Therefore, no pretreatment for removing the toxins in the sludge was tried before enzymatic hydrolysis.

Table I. Effect of water extractable materials on enzyme activities

Dissolving solvent	Sludge washing water (1)	Distilled water (2)	Activities in (1) / Activities in (2)
Enzyme acitivity	U/ml	U/ml	
Cellulase (*Trichoderma reesei*)			
CMCase	15.32	8.98	1.71
Avicelase	0.20	0.22	0.91
FPase	0.64	3.51	0.18
CMCase/FPase	24.12	2.56	
β-glucosidase			
NOVOZYM 188	251	250	≈ 1

The measurements were duplicated.

SOURCE: Reproduced with permission from reference 12. Copyright 2001 The Korean Society for Microbiology and Biotechnology.

Effect of Sludge Pre-drying and Initial Glucose Concentrations on Sludge Hydrolysis

Sludge was dried at 80°C for two days and re-wetted before hydrolysis experiment. Cellulase with CMC activity of 10 U/ml and β-glucosidase activity of 10 U/ml were used in the experiments. A control using non-dried sludge was used in the experiment. In the experiment to see the initial glucose effect on hydrolysis, three levels of initial glucose concentrations of 5, 10 and 20 g/l and a control without addition of glucose were used (*12*). The result of sludge pre-drying effect on hydrolysis was shown in Fig. 1. It showed that 64% decrease of glucose production occurred in case of using re-wetted sludge after drying than using non-dried sludge after 13 h of hydrolysis. The result of initial glucose concentrations on hydrolysis was shown in Fig. 2. It showed that initial glucose concentrations of 5, 10 and 20 g/l resulted in 9.5%, 20.4% and 39.2% decreases, respectively, in glucose production after 16 h of hydrolysis.

The structure of cellulose fibers is much affected by water. Changes in the quantity of hygroscopically bound water govern many physical properties including the extent of swelling and shrinkage. An ultrastructural model of cellulose fibers of non-dried, dried and re-wetted after drying was suggested (*13,14*). It was reported that non-dried cellulose has more surface areas than the re-wetted after drying ones, while the dried cellulose has the least surface areas among the three kinds. Fan, *et al.* (1980) reported that the re-wetted cellulose fibers after drying had 60% less surface area than the non-dried ones (*15*). The

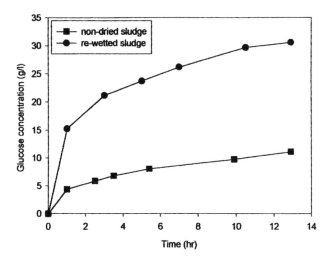

Fig. 1. Time course of hydrolysis of re-wetted and non-dried sludge.

126

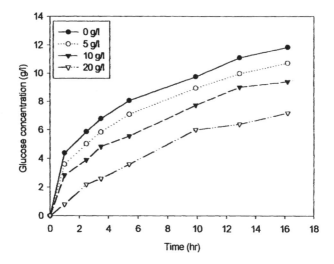

Fig. 2. Time course of glucose production with varing initial glucose concentration.

less glucose yield from the re-wetted sludge after drying than that of the non-dried sludge in our experiment was resulted from the decrease of the surface areas of the cellulose fibers, which are needed for the biding of the cellulase during the hydrolysis.

Glucose, the final product, inhibits various steps in the enzymatic hydrolytic process of cellulose into glucose (10), which leads to obvious decrease in glucose production in our experiment, especially when the initial glucose concentration is higher than 10 g/l (Fig. 2). Therefore, continuous removal/consumption of glucose can increase the productivity of the hydrolytic process.

Optimization of Enzyme Systems for Hydrolysis of Sludge of Varing Concentration

Sludge of 5, 10 and 20% were used in the hydrolysis experiments. Cellulase and β-glucosidase were mixed in various combinations. In hydrolysis of 5% sludge, six levels of CMC activities, 200, 100, 10, 5, 3 and 1 U/ml, were tested. Four levels of β-glucosidase activities, 20, 10, 1 and 0.25 U/ml, were tested.

127

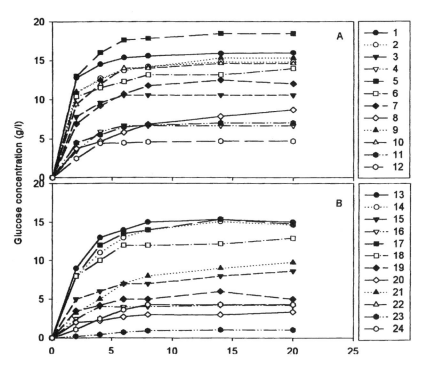

Fig. 3. Time course of 5% sludge hydrolysis. A: CMCase/β-glucosidase activities for case 1 to 12 of 200/20, 200/10, 200/1, 200/0.25, 100/20, 100/10, 100/1, 100/0.25, 10/20, 10/10, 10/1, 10/0.25 U/ml, respectively, B: CMCase/β-glucosidase activities for case 13 to 24 of 5/20, 5/10, 5/1, 5/0.25, 3/20, 3/10, 3/1, 3/0.25, 1/20, 1/10, 1/1, 1/0.25 U/ml, respectively.

Reproduced with permission from reference 12. Copyright 2001 The Korean Society for Microbiology and Biotechnology.

128

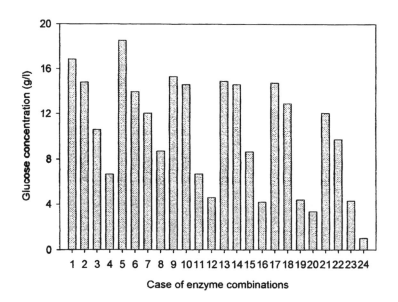

Fig. 4. Hydrolysis of 5% sludge using various enzyme systems. Cases from 1 to 24 are the same as cases in Fig. 3.
Reproduced with permission from reference 12. Copyright 2001 The Korean Society for Microbiology and Biotechnology.

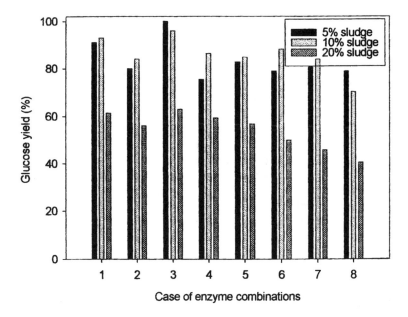

Fig. 5. Glucose yields in sludge hydrolysis. Cases 1 to 8 stand for CMCase/β-glucosidase activities of 200/20, 100/20, 200/10, 100/10, 10/20, 10/10, 5/20 and 5/10 U/ml, respectively.
Reproduced with permission from reference 12. Copyright 2001 The Korean Society for Microbiology and Biotechnology.

All combinations of CMC and β-glucosidase activities were tested for hydrolysis of 5% sludge in a batch mode. The time course of the hydrolysis showed that glucose concentration reached the maximum within 6 hours of hydrolysis (Fig. 3). Glucose production was best in the case 5 of the CMC activity of 100 U/ml and β-glucosidase activity of 20 U/ml (Fig. 4). With the increase of CMC activity to 100 U/ml, the case 1, the glucose production decreased. High glucose productions were obtained at high β-glucosidase activities.

The results showed that CMC activity at about 5 U/ml and β-glucosidase activity at about 10 U/ml was an optimal condition for hydrolysis of 5% sludge, considering the productivity per unit amount of cellulases. Glucose yield was increased with high β-glucosidase activities. With CMC activity of 100 U/ml and β-glucosidase activity of 20 U/ml, glucose yield was almost one hundred percent (Fig. 5).

In general, the higher the enzyme activity, the more glucose was produced. However, with the same β-glucosidase activity of 20 U/ml, glucose production with CMC activity of 200 U/ml was lower than that with 100 U/ml.

The former simulation result showed that with high activities of cellulase, not only glucose but also cellobiose concentration reached a high level in the early stage of the hydrolysis (data not shown). Cellobiose accumulation is known to inhibit the enzymatic activity of both the endo- and exo-type of glucanase components of the fungal cellulase complex (*16*). This might be the reason why too high CMC or cellulase activities lead to the decrease of final glucose concentration.

In hydrolysis of 10% sludge, four levels of CMC activities, 200, 100, 10 and 5 U/ml, respectively, and two levels of β-glucosidase activities, 20 and 10 U/ml, respectively, were used. All combinations of the levels of CMC and β-glucosidase activities were tried in hydrolysis of 10% sludge in the batch mode. Glucose concentration increased quickly and almost reached the maximum level after 10 hours of hydrolysis (Fig. 6).

The highest glucose production was obtained with CMC and β-glucosidase activities of 100 and 20 U/ml, respectively. At the same β-glucosidase activity of 20 U/ml, glucose production with CMC activity of 200 U/ml was lower than that with 100 U/ml(Fig. 7), which was the same pattern as that in hydrolysis of 5% sludge. Glucose production was not changed much in all eight cases of enzyme combinations (Fig. 7). Glucose yields were also high enough and had similar values as in hydrolysis of 5% sludge (Fig. 5). Glucose production almost doubled with sludge of 10% compared to 5% without any indication of an apparent product inhibition effect. The CMC activity of 5 or 10 U/ml and β-glucosidase activity of 10 U/ml was found to be an optimal enzyme combination for hydrolysis of 10 % sludge.

In hydrolysis of 20% sludge, same enzyme combinations and operation conditions as hydrolysis of 10% sludge were used. All combinations of four levels of CMC activities, 200, 100, 10 and 5 U/ml and two levels of β-glucosidase activities, 20 and 10 U/ml were used. In cases of CMC activities of 200 or 100 U/ml, glucose concentration almost reached the maximum level after 10 hours of hydrolysis (Fig. 8). However, in cases of lower CMC activities, such as 10 or 5 U/ml, glucose concentration reached the maximum level after 20 hours of hydrolysis (Fig. 8).

The highest glucose production was also obtained with CMC activity of 100 U/ml and β-glucosidase activity of 20 U/ml, which was the same optimal condition as hydrolysis of sludge of 5 or 10% (Fig. 9). Except the cases of CMC activity of 200 U/ml, glucose production decreased more obviously with the decrease of enzyme activities for both CMCase and β-glucosidase in hydrolysis of sludge of 20% compared to that of 5 or 10% (Fig. 4, 7 and 9).

Glucose yields in hydrolysis of sludge of 20% were apparently lower than that of 5 or 10% (Fig. 5). This was because the product inhibition effect caused by cellobiose and glucose was stronger in case of 20% than that of 5 or 10%. Besides the product inhibition problem, mass transfer resistance also occurred in

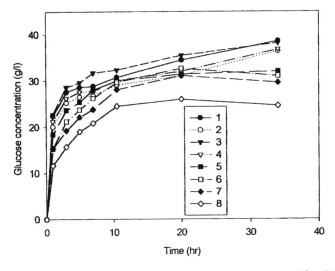

Fig. 6. Time course of 10% sludge hydrolysis. Cases 1 to 8 stand for CMC and
β-glucosidase activities of 200/20, 200/10,100/20, 100/10, 10/20, 10/10,
5/20,5/10, repectively.
Reproduced with permission from reference 12. Copyright 2001 The Korean
Society for Microbiology and Biotechnology.

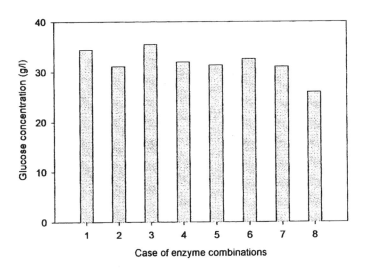

Fig. 7. Results of 20 hour hydrolysis of 10 % sludge with various enzyme
systems. Cases from 1 to 8 are the same as cases in Fig. 6.
Reproduced with permission from reference 12. Copyright 2001 The Korean
Society for Microbiology and Biotechnology.

132

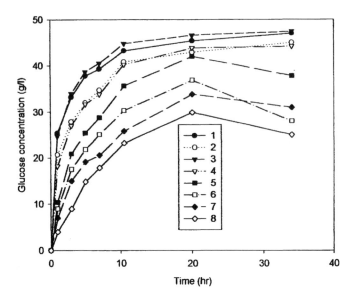

Fig. 8. Time course of 20 % sludge hydrolysis. Cases from 1 to 8 are the same
as cases in Fig. 6.
Reproduced with permission from reference 12. Copyright 2001 The Korean
Society for Microbiology and Biotechnology.

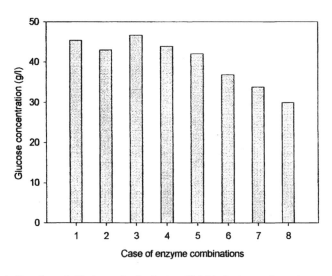

Fig. 9. Results of 20 hour hydrolysis of 20 % sludge with various enzyme
systems. Cases from 1 to 8 are the same as cases in Fig. 6.
Reproduced with permission from reference 12. Copyright 2001 The Korean
Society for Microbiology and Biotechnology.

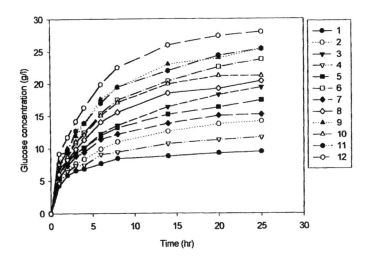

Fig. 10. Time course of hydrolysis with intermittent feeds of sludge. At 0, 1, 2, 3 h of hydrolysis, 5% sterilized sludge was added to make total sludge amount of 20%. Cases 1 to 12 stand for CMCase/β-glucosidase activities are 5/2.5, 5/5, 5/10, 10/2.5, 10/5, 10/10, 20/2.5, 20/5, 20/10, 50/2.5. 50/5 and 50/10 U/ml, respectively.
Reproduced with permission from reference 12. Copyright 2001 The Korean Society for Microbiology and Biotechnology.

case of 20%, especially with lower enzyme activities. Intermittent feeding of sludge could not increase either glucose production or glucose yield even though it could overcome the mass transfer problem (Fig. 10, 11 and 12). The reason was that in case of product inhibition, high initial concentration of substrate favored the reaction. CMCase activity of 10 U/ml and β-glucosidase activity of 20 U/ml was an optimal enzyme combination for hydrolysis of sludge of 20% in batch mode. In order to develop a process of practical importance, crude cellulase of fixed endo and exo-type glucanase composition was used. Therefore, no optimization on the composition of endo and exo-type glucanase was made in this research.

High glucose yield can be obtained by enzymatic hydrolysis especially at sludge concentrations lower than 20%. This result shows that the sludge needs no pretreatment before enzymatic hydrolysis in order to obtain high glucose yield. Besides the chemicals that are added in the paper mill, the properties of the cellulosic materials such as particle size, degree of polymerization (DP), and crystallinity index (CrI) also affect the enzymatic hydrolysis of paper mill sludge.

134

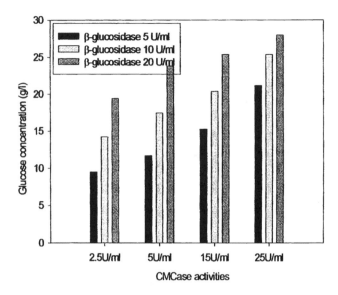

Fig. 11. Glucose concentrations at 25 hour of hydrolysis with intermittent feeds
of sludge.
Reproduced with permission from reference 12. Copyright 2001 The Korean
Society for Microbiology and Biotechnology.

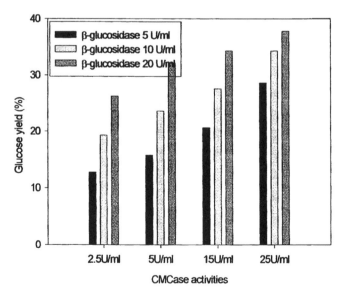

Fig. 12. Glucose yields of sludge hydrolysis in fed-batch mode.
Reproduced with permission from reference 12. Copyright 2001 The Korean
Society for Microbiology and Biotechnology.

Cellulose is usually present in a bundle of fibrillar units with a supramolecular structure consisting of crystalline and amorphous regions (*17*). The amorphous regions of the fiber are supposed to occur at similar intervals within the fiber. The initial action of the enzyme is to attack the less ordered regions of the fiber resulting in a rapid reduction in particle size. Rapid decrease in viscosity of the sludge during the hydrolysis process was observed. But, the influences of DP and crystallinity in restricting the ease of enzymatic hydrolysis are unclear. The susceptibility of sludge to complete enzymatic hydrolysis can not be easily predicted from the differences in the initial DP and crystallinity (*18,19*). Therefore, DP and crystallinity were not measured in this research.

Enzyme related factors that are influential in the hydrolysis process include end product inhibition of the enzyme, thermal and mechanical inactivation, and irreversible adsorption of the enzymes (*20,21,22*). In this research, glucose production and the glucose yield were increased by the optimization of the enzyme system through the addition of β-glucosidase, which hydrolyzes cellobiose to glucose, thereby preventing inhibition of cellulase by cellobiose (*23*). Optimization of the enzyme system can also be made using mixed culture of *Trichoderma reesei* and *Aspergillus niger* (*24,25,26*), which are the efficient organisms producing cellulase and β-glucosidase, respectively. The mixed culture is being done in our laboratory in order to produce the optimized enzyme system more efficiently. Thermal and mechanical inactivation should not be severe at the temperature not higher than 50°C and the agitation speed lower than 100 rpm during the reaction period within two days, which condition is going to be used in our later research. Lignin is reported to be responsible for irreversible adsorption of cellulases (*27,28,29*), but there is no lignin in the sludge that we use.

The glucose concentration of about 30 g/l, produced by hydrolysis of 10% sludge is high enough for a fermentation process operated in a continuous or repeated fed-batch mode. In order to make the hydrolysis process more economical, we are using sludge as the material to produce the enzymes in low cost after confirming that *Trichoderma reesei* and *Aspergillus niger* can grow and produce enzymes normally on sludge. Simultaneous saccharification and fermentation (SSF) process can improve the hydrolysis efficiency by prevention of the accumulation of glucose and cellobiose to reduce the severe feedback inhibition effect on hydrolytic enzymes (*11,29,30*). It has been reported that lactic acid, acetic acid, citric acid, itaconic acid, α-ketoglutaric acid, and succinic acid scarcely inhibit cellulase (*31*). As a result, SSF has the advantage in producing the organic acids using cellulosic materials. We have confirmed that conditions for L-lactic acid fermentation using *Lactobacillus* and those for cellulase and β-glucosidase reaction are compatible, which makes it possible to produce L-lactic acid from paper mill sludge using SSF process. This process is hopefully to reduce the cost of L-lactic acid, low enough for making degradable plastic (*32*).

Acknowledgements

This research was supported by ERC for Advanced Bioseparation Technology and Visiting Scholar Foundation of Key Laboratory in University, Ministry of Education of China.

REFERENCES

1. Ooshima, H.; Sakata, M.; Harano, Y. Adsorption of cellulase from *Trichoderma viride* on cellulose. *Biotechnol. Bioeng.* **1983**, *25*, 3103-3114.
2. Withers, S. G.; Tull, D.; Gebler, J.; Braun, C.; Aebersold, R.; Wang, Q.; Warren, T.; Kilburn, D.; Gilkes, N.; Suominen, P.; Reinikainen, T. *Mechanistic studies on cellulases. In: Proceedings of the second TRICEL symposium on Trichoderma reesei cellulases and other hydrolases;* Eds.; Espoo, Finland, 1993; pp 117-123.
3. Nazhad, M.; Ramos, M.; Paszner, L. P. L.; Saddler, J. N. Structural constrains affecting the innitial enzymatic hydrolysis of recycled paper. *Enzyme Microb. Technol.* **1995**, *17*, 68-74.
4. Eklund, R.; Galbe, M.; Zacchi, G. Optimization of temperature and enzyme concentration in the enzymatic saccharification of steam pretreated willow. *Enzyme Microb. Technol.* **1990**, *12*, 225-228.
5. Philippidis, G. P.; Smith, T. K.; Wyman, C. E. Study of the enzymatic hydrolysis of cellulose for production of fuel ethanol by the simultaneous saccharification and fermentation process. *Biotechnol. Bioeng.* **1993**, *41*, 846-853.
6. Ghosh, T. K.; Das, K. Kinetics of hydrolysis of insoluble cellulose by cellulase. *Adv. Biochem. Eng.* **1971**, *1*, 49-55.
7. Mansfield, S. D.; Mooney, C.; Saddler, J. N. Substrate and enzyme characteristics that limit cellulose hydrolysis. *Biotechnol. Prog.* **1999**, *15*, 804-816.
8. Ooshima, H.; Burns, D. S.; Converse, A. O. Adsorption of cellulase from *Trichoderma reesei* on cellulose and lignacious residue in wood pretreated by dilute sulfuric acid with explosive decompression. *Biotechnol. Bioeng.* **1990**, *36*, 446-452.
9. Ghose, T. K.; Panda, T.; Visaria, V. S. Effect of culture phasing and mannanase on production of cellulase and hemicellulase by mixed culture

of *Trichoderma reesei* D1-6 and *Aspergillus wentii* Pt 2804. *Biotechnol. Bioeng.* **1985**, *27*, 1353-1361.

10. Parajo, J. C.; Alonso, J. L.; Santos. V. Development of a generalized phenomenological model describing the kinetics of the enzymatic hydrolysis of NaOH-treated pine wood. *Applied Biochem. Biotechnol.* **1996**, *56*, 289-299.

11. Moritz, J. W.; Duff, S. J. B. Simultaneous saccharification and extravtive fermentation of cellulosic substrates. *Biotechnol. Bioeng.* **1995**, *49*, 504-511.

12. Lin, J.; Lee, S. M.; Koo, Y. M. Hydrolysis of paper mill sludge using an improved enzyme system. *J. Mirobiol. Biotech.* **2001**, *11*, 362-368.

13. Duff, S. J. B.; Cooper, D. G.; Fuller. O. M. Cellulase and β-glucosidase production by mixed culture of *Trichoderma reesei* RUT C30 and *Aspergillus phoenicis*. *Biotech. Lett.* **1985**, *7*, 185-190.

14. Fan, L. T.; Lee, Y. H.; David, H. Mechanism of the enzymatic hydrolysis of celluloses: effects of major structural features of cellulose on enzymatic hydrolysis. *Biotechnol. Bioeng.* **1980**, *XXII*, 177-199.

15. Nakasaki, K.; Akakura, N.; Adachi, T.; Akiyama, T. Use of wastewater sludge as a raw material for production of L-lactic acid. *Environ. Sci. Technol.* **1999**, *33*, 198-200.

16. Breuil, C.; Chan, M.; Gilbert, M.; Saddler. J. N. Influence of β-glucosidase on the filter paper activity and hydrolysis of lignocellulosic substrates. *Bioresourse Technol.* **1992**, *39*, 139-142.

17. Aabe, S. I.; Takagi, M. Simultaneous saccharification and fermentation of cellulose to lactic acid. *Biotechnol. Bioeng.* **1991**, *37*, 93-96.

18. Cleresci, L. S.; Sinitsyn, A. P.; Saunders, A. M.; Bungay, H. R. Recycle of cellulase enzyme complex after hydrolysis of steam-exploded wood. *Applied Biochem. Biotechnol.* **1985**, *11*, 433-443.

19. Ladisch, M. R.; Svarczkopf, J. A. Ethanol production and the cost of fermentable sugars from biomass. *Bioresourse Technol.* **1991**, *36*, 83-95.

20. Saddler, J. N. Factors limiting the efficiency of cellulase enzymes. *Microbiol. Sci.* **1986**, *3*, 84-87.

21. Scallan, A. M. Fibre-water interactions in paper-making. In: Trans Symp Oxford, Fundamental Research Committee (ed) Tech. Div. Brit. PaperBoard Ind. Fed., London, 1978, *1*, 9-27.

22. Sternberg, D. Production of cellulase by *Trichoderma*. *Biotechnol. Bioeng.* **1976**, *6*, 35-53.

23. Coughlan, M. P. Enzymatic hydrolysis of cellulose: an overview. *Bioresourse Technol.* **1992**, *39*, 107-115.

24. Panda, T.; Visaria, V. S.; Ghose, T. K. Studies on mixed fungal culture for cellulase and hemicellulase production. Part 1: Optimization of medium for

the mixed culture of *Trichoderma reesei* D1-6 and *Aspergillus wentii* Pt 2804. *Biotech. Lett.* **1983**, *5*, 767-772.

25. Sutcliffe, R.; Saddler, J. N. The role of lignin in the adsorption of cellulases during enzymatic treatment of lignocellulosic material. *Biotechnol. Bioeng.* **1986**, *25*, 256-261.

26. Duff, S. J. B.; Moritz, J. W.; Andersen. K. L. Simultaneous hydrolysis and fermentation of pulp mill primary clarifier sludge. *Can. J. Chem.Eng.* **1994**, *72*, 1013-1020.

27. Kaur, P. P.; Arneja, J.` S.; Singh, J. Enzymatic hydrolysis of rice straw by crude cellulase from *Trichoderma reesei. Bioresourse Technol.* **1998**, *66*, 267-269.

28. Ooshima, H.; Kurakake, M.; Kato, J.; Harano. Y. Enzymatic activity of cellulase adsorbed on cellulose and its change during hydrolysis. *Appl. Biochem. Biotechnol.* **1991**, *31*, 253-266.

29. Stockton, B. C.; Grohmann, D. J. K.; Himmel, M. E. Optimum β-glucosidase supplementation of cellulase for efficient conversion of cellulose to glucose. *Biotech. Lett.* **1991**, *13*, 57-62.

30. Yu, X.; Yun, H. S.; Koo, Y. M. Production of cellulase by *Trichoderma reesei* Rut C30 in a batch fermenter. *J. Microbiol. and Biotechnol.* **1998**, *8*, 575-580.

31. Moritz, J. W.; Duff, S. J. B. Ethanol production from spent sulfite liquor fortified by hydrolysis of pulp mill primary clarified sludge. *Applied Biochem. Biotechnol.* **1996**, *57/58*, 689-698.

32. Gritzali, M.; Brown, R. D. The cellulase system of *Trichoderma.* In: Hydrolysis of cellulose: Mechanisms of enzymatic and acidic catalysis. R.D. Brown and L. Jurasek, eds., American Chemical Society: Washington, DC, 1979; *181*, pp 237-260.

Animal Cell Systems

Chapter 11

Antigen-Mediated Genetically Modified Cell Amplification

Masahiro Kawahara[1], Hiroshi Ueda[1,2,*], Kouhei Tsumoto[3], Izumi Kumagai[3], Walt Mahoney[4], and Teruyuki Nagamune[1]

[1]Department of Chemistry and Biotechnology, Graduate School of Engineering, The University of Tokyo, Hongo, Bunkyo-ku, Tokyo 113–8656, Japan
[2]Department of Integrated Biosciences, Graduate School of Frontier Sciences, The University of Tokyo, Kashiwanoha, Kashiwa 277–8562, Japan
[3]Department of Biomolecular Engineering, Graduate School of Engineering, Tohoku University, Aoba, Aramaki, Aoba-ku, Sendai 980–8579, Japan
[4]Quantum Dot Corporation, 26136 Research Road, Hayward, CA 94545

Hematopoietic stem cells (HSCs), due to their pluripotency, are the major targets for gene therapy focused on correcting hematopoietic diseases. However, in most cases poor transduction efficiency of the therapeutic gene into HSCs has led to only modest clinical improvement. To overcome this shortcoming we propose a selective cellular expansion method using an antibody/receptor chimera trigger that can be activated by a monomeric antigen. Two plasmids encoding 1) a hybrid receptor composed of the V_H portion of anti-hen egg lysozyme antibody HyHEL-10 and a N-terminally truncated erythropoietin receptor (V_H-EpoR), and 2) a V_L-EpoR fusion derived from the same construct as in 1, were employed. The second plasmid also encoded enhanced green fluorescent protein (EGFP) as a model therapeutic gene that was flanked

© 2002 American Chemical Society

by the internal ribosomal entry sequence. Both plasmids were used simultaneously to transfect an IL-3 dependent murine myeloid cell line, 32D. The transfectants after antigen selection in the absence of IL-3 showed a clear antigen induced dose-dependent proliferation. In addition, by flow cytometry higher EGFP expression level was observed than that of the cells before antigen selection. The results demonstrate the expansion of genetically modified hematopoietic cells *in vitro* and possibly *in vivo*. We propose the term AMEGA (Antigen MEdiated Genetically-modified cell Amplification) for such an approach.

Introduction

Gene therapy holds great promise as an effective therapeutic modality to correct many inherited and acquired diseases. While results to date have had mixed success, clinical trials employing gene therapy are underway for such diverse conditions as adenosine deaminase (ADA) deficiency (*1*), Fanconi anemia (*2*), β-thalassemia (*3*), cancer (*4*), acquired immunodeficiency syndrome (AIDS) (*5*) and many others. To attain a therapeutic effect sufficient to treat hematopoietic diseases, efficient gene transduction into hematopoietic stem cells (HSCs) is critical. However, transduction efficiency is not adequate at the moment in spite of intensive efforts to improve vectors (*6-9*). Retroviral vectors derived from oncoretroviruses have been widely used for gene transduction into HSCs. However, since retroviral genes cannot be integrated into non-dividing cells, the transduction efficiency for quiescent HSCs is low (*10-12*). Furthermore, in previous clinical trials gene-transduced cells were merely injected into the patients' body and no efforts were made for maintaining or selecting the transduced cells *in vivo*. In the cases of ADA deficiency and Fanconi anemia, growth advantage of transduced cells could contribute to an enhanced therapeutic effect, but in other diseases such as Gaucher disease (*13*) and chronic granulomatous disease (*14*), a growth advantage of transduced cells is not expected.

To improve the therapeutic efficiency for such diseases, selective expansion of gene-transduced cells has been considered (*15-19*). Recently, a modified receptor gene (truncated erythropoietin receptor: tEpoR) has been proposed as an improved gene amplifier (*20,21*). With these methods, cells with transgenes are expected not only to survive but also to proliferate, which is a major advantage in contrast to conventional drug resistance selections. However, side effects promoted by the administration of cytokines or other bioactive substances can be also problematic in the case of *in vivo* administration.

To solve this problem, we propose to use an artificial receptor recognizing a ligand that has no toxic effect on normal cells. Previously, we showed that the interchain interaction between the variable regions (V_H and V_L) of anti-hen egg lysozyme (HEL) antibody HyHEL-10 (*22*) is very weak (Ka < 10^5) in the absence of HEL, whereas the interaction becomes strong (Ka ~ 10^9) in the presence of HEL (*23*). Then we reasoned that if the V_H or V_L region of HyHEL-10 is substituted for the ligand-binding domain of certain cytokine receptor, we could create a novel receptor that could be activated by a specific antigen, HEL. More specifically, the N-terminal part (D1) of the extracellular domain of EpoR was replaced with one of the variable regions (V_H or V_L) of HyHEL-10, which here we call HE or LE chains, respectively. The co-transfection of two plasmids encoding HE and LE chains to IL-3 dependent hematopoietic cells resulted in efficient cell growth signal transduction in response to antigen HEL (*24*). Here we try to apply this technology for enhancing the therapeutic gene expression in the hematopoietic cells. The enhanced green fluorescent protein (EGFP) gene was employed as a model transgene under the control of the internal ribosomal entry site (IRES) (*25*) downstream of the LE gene to create LE-IRES-EGFP gene. HE and LE-IRES-EGFP genes were simultaneously transfected into a hematopoietic cell line to see if the antigen selection could successfully expand the EGFP-positive cells.

Materials and Methods

Plasmid Construction

pME-VHER and pME-VLER were made from pME-ER by replacing EpoR extracellular D1 domain with V_H and V_L region of HyHEL-10, respectively, with no linker sequence between V_H/V_L and EpoR (*24*). The 'amplifier' plasmid pME-HE, which is an expression vector for the HE chain encoding HyHEL-10 V_H, GSG linker, D2, transmembrane and cytoplasmic domains of EpoR, was previously described as pME-VHER(GSG) (*24*). The plasmid pMEZ-LE, which is an expression vector for the LE chain, is a zeocin resistance variant of previously described pME-VkER(GSG). To make the 'courier' plasmid pMEZ-LEIGFP, pIRES2-EGFP (Clontech, Palo Alto, CA) was digested with *Sma* I, and *Xba* I linker (5'-CTCTAGAG-3') was ligated to make a *Xba* I site (underlined) at the upstream of the IRES-EGFP gene, resulting in pIGFP-X. pIGFP-X was digested with *Xba* I to obtain the IRES-EGFP fragment, which was subsequently ligated to the *Xba* I-digested pMEZ-LE to give pMEZ-LEIGFP which encodes IRES-EGFP at the downstream of LE chain.

Cell Culture

A murine IL-3-dependent pro-B cell line Ba/F3 (*26*) and a murine IL-3-dependent myeloid cell line 32D (*27,28*) were cultured in RPMI 1640 medium (Nissui Pharmaceutical, Tokyo) supplemented with 10 % FBS (Iwaki, Tokyo) and 2 ng/ml murine IL-3 (Genzyme, Cambridge, MA) at 37 °C in a 5 % CO_2 incubator.

Transfection of the Plasmids

Transfection of Ba/F3 cells was as described previously (*24*). 32D cells (3 × 10^6 cells) were washed and resuspended with 500 µl of Hanks' buffered saline (Nissui Pharmaceutical), and mixed with 10 µg each of pME-HE and pMEZ-LEIGFP. The mixture was transferred to a 4 mm-gapped electroporation cuvette, and after a 10 min incubation at room temperature, the mixture was electroporated with an Electroporator II (Invitrogen, Groningen, Netherlands) set at 250 µF, 600 V. After a 10 min incubation at room temperature, the cells were transferred to 10 ml of medium in a φ100 mm culture dish and incubated at 37 °C for 2 days, followed by selection with 480 µg/ml G418 (Sigma, St. Louis, MO) and 500 µg/ml zeocin (Invitrogen). The antibiotic-resistant cells (32D/HE+LEIGFP (3GZ)) were further selected in the medium without IL-3 but with 10 µg/ml HEL (Seikagaku Corporation, Tokyo). The survived cells were named 3GZH.

Western Blotting

The cells (10^6 cells) were washed with PBS, lysed with 100 µl lysis buffer (20 mM HEPES (pH 7.5), 150 mM NaCl, 10 % Glycerol, 1 % Triton X-100, 1.5 mM $MgCl_2$, 1 mM EGTA, 10 µg/ml aprotinin, 10 µg/ml leupeptin) and incubated on ice for 10 min. After centrifugation at 15 krpm for 5 min, the supernatant was mixed with Laemmli's sample buffer and boiled. The lysate was resolved by SDS-PAGE and transferred to a nitrocellulose membrane (Millipore, Bedford, MA). After the membrane was blocked with 5% skim milk, the blot was probed with rabbit anti-mouse EpoR antibody (Santa Cruz Biotechnology, Santa Cruz, CA) followed by HRP-conjugated anti-rabbit IgG

(Biosource, Camarillo, CA), and detected with ECL system (Amersham-Pharmacia, Buckinghamshire, UK).

Cell Proliferation Assay

Transfectant cells were washed three times with PBS and seeded in 24- or 96-well plates containing RPMI1640 medium supplemented with 10 % FBS and various concentration of HEL. Cell number was counted with a hemocytometer, or estimated with WST-1 assay (*29*).

Flow Cytometry

Cells were washed with PBS and resuspended in Isoflow (Beckman Coulter, Fullerton, CA). Green fluorescence intensity was measured with EPICS-C (CS) flow cytometry (Beckman Coulter) at 488 nm excitation and fluorescence detection at 525 ± 20 nm.

Results

Ba/F3 Transfectant Showed Antigen-Dependent Proliferation

We first employed IL-3 dependent pro-B cell line Ba/F3, which was shown to transduce a growth signal in response to Epo when wild type EpoR was ectopically expressed. The cells were electroporated with pME-VHER and pME-VLER with no linker sequence between V_H/V_L and EpoR. After antibiotic selection, the survived cells were further selected in $HEL^+IL\text{-}3^-$ medium, resulting in HEL-dependently proliferating colonies. The selected cells were cloned and named Ba/HE+LE cells.

Ba/HE+LE cells expressed both HE and LE chains, which were confirmed by Western blot with anti-EpoR antibody. The affinity purification of cell lysate with streptavidin-agarose and subsequent Western blot with anti-EpoR antibody revealed the biotinylated HEL dose-dependent increase of band density corresponding to HE and LE chains, indicating that HEL-dependent association of HE and LE chains. Cell proliferation assay showed HEL concentration-dependent proliferation of Ba/HE+LE cell clones, suggesting that the chimeric receptors were functional for signal transduction (Figure 1). The growth rate of Ba/HE+LE cells was similar to that of Ba/EpoR cells, which are Ba/F3 transfectants with wild-type EpoR. STAT5b, which is one of the transcription factors involved in EpoR-mediated signaling, was phosphorylated in response to HEL, further confirming that chimeric receptor mimicked wild-type EpoR-derived signal transduction. However, significant background cell growth was observed without HEL addition, implying the requirement to improve the receptor construct to elicit a stricter antigen-dependent proliferation switch.

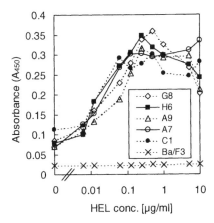

Figure 1. HEL-dependent growth of Ba/HE+LE cell clones on day 6. Initial cell concentration was 5×10^3 cells/ml.

Linker Insertion between Fv and EpoR

We speculated that the linker addition might change the growth characteristics through conformational effect as suggested by crystallographic analyses. Therefore, we inserted Gly (G) or Gly-Ser-Gly (GSG) linker between V_H (or V_L) and EpoR to modify the flexibility of the chimeric receptors. The pairs of constructs with G linker, GSG linker, and no linker were transfected to IL-3-dependent myeloid cell line 32D by electroporation, followed by G418 and HEL selection to create 32D/G, and 32D/N cells, respectively. While 32D/G and 32D/N cells grew significantly in IL-3⁻HEL⁻ medium, 32D/GSG cells showed clear HEL dose-dependent growth with undetectable background in IL-3⁻HEL⁻ medium (Figure 2). Hence we used 32D/GSG cells for further study.

Selective Expansion of 32D Transfectants with Higher EGFP Expression Level

To show the possibility for selective amplification of transgene-expressing cells, EGFP gene, flanked by internal ribosomal entry site (IRES), was inserted at the downstream of LE gene as a model therapeutic gene, resulting in pMEZ-LEIGFP. 32D cells were electroporated with G418 resistant pME-HE and zeocin resistant pMEZ-LEIGFP by electroporation, followed by selection with simultaneous addition of G418 and zeocin. Flow cytometric analysis revealed that the antibiotic-resistant cells (named 3GZ cells) were a mixture of cell populations with a wide range of EGFP expression levels (Figure 3). These cells were further selected in IL-3⁻HEL⁺ medium, resulting in a portion of survived cells. The HEL-selected cells (named 3GZH cells) showed a single green fluorescent peak corresponding to the highest green fluorescent peak observed in antibiotic-resistant cells (Figure 3). These results indicate that HEL induced specific amplification of the cell population with the highest EGFP expression level.

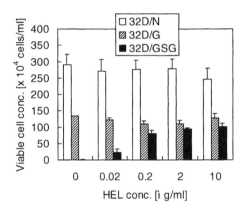

Figure 2. HEL-dependent growth of 32D/N (day 4; initial cell concentration was 7.7 × 10⁴ cells/ml), 32D/G (day 6.8; 1.2 × 10⁴ cells/ml) and 32D/GSG (day 6.8; 1.1 × 10⁴ cells/ml).

HEL-Dependent Cell Growth of Cells after HEL Selection

To confirm that HEL specifically induced growth stimulation of EGFP-positive cells, a cell proliferation assay was performed. As shown in Figure 4, HEL induced cell proliferation in a dose-dependent manner, whereas cells died without HEL addition. More than 100 ng/ml HEL supported cell viability of more than 70 % during the assay period. 10 ng/ml HEL was the lower limit for the growth maintenance, although some anti-apoptotic effect was observed with addition of 1 ng/ml HEL. This lower concentration limit was comparable to the amount required for cell proliferation of 32D/GSG cells (Figure 2). Additionally, this level roughly corresponds to the equilibrium dissociation constant K_d (1nM) of the complex composed of V_H, V_L and HEL (24). These results indicate that HEL induced specific amplification of EGFP-positive cells by transducing efficient cell growth and anti-apoptotic signals.

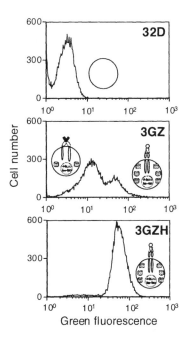

Figure 3. Flow cytometric analysis of parental 32D, 32D/HE+LEIGFP (3GZ), and 32D/HE+LEIGFP (3GZH).

Correlation between Chimeric Receptor and EGFP Expression Levels

To compare the amounts of expressed chimeric receptor chains, cell lysates from 3GZ and 3GZH cells were prepared and the expression of HE and LE chains was detected using Western blotting with an anti-EpoR antibody (Figure 5). Both HE and LE chains were expressed in 3GZ and 3GZH cells, but the corresponding levels of the HE and LE chains differed between the two cell types. 3GZH cells showed consistently higher LE chain expression than 3GZ cells. This is quite reasonable from the fact that EGFP expression level was

148

expected to correlate with LE chain expression level since the LE gene locates
upstream of the IRES-EGFP gene, and that cells with higher EGFP expression
level were specifically amplified after antigen selection.   As a result, the
amplified cells were controllable with the antigen, while keeping elevated
expression of EGFP transgene.

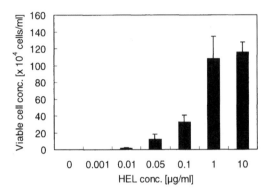

*Figure 4. HEL-dependent growth of 3GZH cells on day 11. Initial cell
concentration was 10³ cells/ml.*

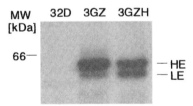

*Figure 5. Altered expression levels of two chimeric receptor chains in 3GZ and
3GZH cells.*

Discussion

In this study, we successfully created a novel chimeric receptor useful for
selective expansion of transfectants with high expression level of a transgene.
Among three 32D transfectants with constructs without linker, or with G or GSG

linkers between V_H/V_L and EpoR domains, 32D/N and 32D/G cells showed nearly constitutive growth without need for HEL. This growth characteristic suggests that the chimeric receptors with G or no linker form active dimers even without HEL addition. On the other hand, 32D/GSG cells exhibited a clear HEL-dependent cell growth, indicating that GSG linker insertion hindered the chimeric receptor from the active state in the absence of HEL. This apparent difference in growth characteristics of three transfectants suggests that the linker insertion altered the amount of preformed dimers, and/or induced their conformational change as shown in previous reports (*30-32*).

Since gene transduction efficiency in HSCs is insufficient at present, various efforts especially including vector improvements have been attempted. Retroviral vectors derived from oncoretroviruses have been widely used for the transfection of hematopoietic cells, and many have been trying to develop high-titer retroviruses. A fibronectin fragment has been developed to improve the transduction efficiency by co-localizing retroviral particles and target cells (*33,34*), but these approaches did not solve the problem of integration efficiency of a transgene. Recently, a human immunodeficiency virus (HIV) vector has been developed which shows promise for efficient transduction to non-dividing cells including HSCs (*35*).

While these approaches were focused on gene transduction efficiency, of utmost importance is the improvement of therapeutic effect. In this context, if transduced cells can be selectively expanded, the therapeutic effect should be improved even if transduction efficiency is low. Based on this premise, selective expansion of transduced cells by growth stimulation has been proposed. In this method, receptor gene as well as target gene is introduced into cells and an exogenously added ligand stimulates the selective proliferation of the transduced cells. However, administration of natural ligands such as erythropoietin and estrogen might result in over proliferation or undesirable response of normal cells. In the case of synthetic ligands such as FK1012 and 4-hydroxytamoxifen, the effects of these ligands on the normal cells *in vivo* are unclear. In our AMEGA (Antigen MEdiated Genetically-modified cell Amplification) system, the number of antigen-antibody pairs is virtually infinite. This means that we can in principle select appropriate ligands from a large pool of candidates suitable for *in vivo* administration. Variation of ligands will not be limited unlike the FKBP- or ER-receptor system where intensive efforts will be required to seek other ligands to bind FKBP or ER, which is intact or mutated (*36,37*). The expanded ligand choice may lead to the possibility of concurrent inputs of distinct signals or different signal outputs such as cell proliferation and cell death signals at different time points. Haptens are the most promising nominees since they also can evade immune response upon *in vivo* administration. Further improvement of this model system by adopting hapten-responsive human chimeric receptors and therapeutic genes will definitely provide *in vivo* application of this system. Since at least one Fv is known to be stabilized by haptens (*38*), more than one system suitable for this purpose will be certainly available in near future.

Acknowledgments

We are grateful to Dr. K. Todokoro (RIKEN) for kindly providing murine EpoR expression vectors. This work was supported by Grant-in-Aids for Scientific Research on Priority Areas (No. 296-10145107), and for Scientific Research (B11555216) from the Ministry of Education, Science, Sports and Culture and funded by Biodesign Research Promotion Group of the Institute of Physical and Chemical Research (RIKEN), Japan.

References

1. Pollok, K. E.; Hanenberg, H.; Noblitt, T. W.; Schroeder, W. L.; Kato, I.; Emanuel, D.; Williams, D. A. *J. Virol.* **1998**, *72*, 4882-4892.
2. Liu, J. M.; Young, N. S.; Walsh, C. E.; Cottler-Fox, M.; Carter, C.; Dunbar, C.; Barrett, A. J.; Emmons, R. *Hum. Gene Ther.* **1997**, *8*, 1715-1730.
3. May, C.; Rivella, S.; Callegari, J.; Heller, G.; Gaensler, K. M. L.; Luzzatto, L.; Sadelain, M. *Nature* **2000**, *406*, 82-86.
4. Tahara, H.; Zitvogel, L.; Storkus, W. J.; Zeh, H. J., 3rd; McKinney, T. G.; Schreiber, R. D.; Gubler, U.; Robbins, P. D.; Lotze, M. T. *J. Immunol.* **1995**, *154*, 6466-6474.
5. Su, L.; Lee, R.; Bonyhadi, M.; Matsuzaki, H.; Forestell, S.; Escaich, S.; Bohnlein, E.; Kaneshima, H. *Blood* **1997**, *89*, 2283-2290.
6. Kume, A.; Hanazono, Y.; Mizukami, H.; Urabe, M.; Ozawa, K. *Int. J. Hematol.* **1999**, *69*, 227-233.
7. Povey, J.; Weeratunge, N.; Marden, C.; Sehgal, A.; Thrasher, A.; Casimir, C. *Blood* **1998**, *92*, 4080-4089.
8. Havenga, M.; Hoogerbrugge, P.; Valerio, D.; van Es, H. H. *Stem Cells* **1997**, *15*, 162-179.
9. Chu, P.; Lutzko, C.; Stewart, A. K.; Dube, I. D. *J. Mol. Med.* **1998**, *76*, 184-192.
10. Miller, D. G.; Adam, M. A.; Miller, A. D. *Mol. Cell. Biol.* **1990**, *10*, 4239-4242.
11. Roe, T.; Reynolds, T. C.; Yu, G.; Brown, P. O. *EMBO J.* **1993**, *12*, 2099-2108.
12. Berardi, A. C.; Wang, A.; Levine, J. D.; Lopez, P.; Scadden, D. T. *Science* **1995**, *267*, 104-108.

13. Migita, M.; Medin, J. A.; Pawliuk, R.; Jacobson, S.; Nagle, J. W.; Anderson, S.; Amiri, M.; Humphries, R. K.; Karlsson, S. *Proc. Natl. Acad. Sci. USA* **1995**, *92*, 12075-12079.

14. Kume, A.; Dinauer, M. C. *J. Lab. Clin. Med.* **2000**, *135*, 122-128.

15. Blau, C. A.; Peterson, K. R.; Drachman, J. G.; Spencer, D. M. *Proc. Natl. Acad. Sci. USA* **1997**, *94*, 3076-3081.

16. Ito, K.; Ueda, Y.; Kokubun, M.; Urabe, M.; Inaba, T.; Mano, H.; Hamada, H.; Kitamura, T.; Mizoguchi, H.; Sakata, T. *Blood* **1997**, *90*, 3884-3892.

17. Jin, L.; Asano, H.; Blau, C. A. *Blood* **1998**, *91*, 890-897.

18. Kume, A.; Ito, K.; Ueda, Y.; Hasegawa, M.; Urabe, M.; Mano, H.; Ozawa, K. *Biochem. Biophys. Res. Commun.* **1999**, *260*, 9-12.

19. Matsuda, K.; Kume, A.; Ueda, Y.; Urabe, M.; Hasegawa, M.; K, O. *Gene Ther.* **1999**, *6*, 1038-1044.

20. Kirby, S. L.; Cook, D. N.; Walton, W.; Smithies, O. *Proc. Natl. Acad. Sci. USA* **1996**, *93*, 9402-9407.

21. Kirby, S.; Walton, W.; Smithies, O. *Blood* **2000**, *95*, 3710-3715.

22. Padlan, E. A.; Silverton, E. W.; Sheriff, S.; Cohen, G. H.; Smith-Gill, S. J.; Davies, D. R. *Proc. Natl. Acad. Sci. USA* **1989**, *86*, 5938-5942.

23. Ueda, H.; Tsumoto, K.; Kubota, K.; Suzuki, E.; Nagamune, T.; Nishimura, H.; Schueler, P. A.; Winter, G.; Kumagai, I.; Mohoney, W. C. *Nat. Biotechnol.* **1996**, *14*, 1714-1718.

24. Ueda, H.; Kawahara, M.; Aburatani, T.; Tsumoto, K.; Todokoro, K.; Suzuki, E.; Nishimura, H.; Schueler, P. A.; Winter, G.; Mahoney, W. C. *J. Immunol. Methods* **2000**, *241*, 159-170.

25. Ghattas, I. R.; Sanes, J. R.; Majors, J. E. *Mol. Cell. Biol.* **1991**, *11*, 5848-5859.

26. Palacios, R.; Steinmetz, M. *Cell* **1985**, *41*, 727-734.

27. Greenberger, J. S.; Eckner, R. J.; Ostertag, W.; Colletta, G.; Boschetti, S.; Nagasawa, H.; Karpas, A.; Weichselbaum, R. R.; Moloney, W. C. *Virology* **1980**, *105*, 425-435.

28. Greenberger, J. S.; Sakakeeny, M. A.; Humphries, R. K.; Eaves, C. J.; Eckner, R. J. *Proc. Natl. Acad. Sci. USA* **1983**, *80*, 2931-2935.

29. Ishiyama, M.; Shiga, M.; Sasamoto, K.; Mizoguchi, M.; He, P. G. *Chem. Pharm. Bull.* **1993**, *41*, 1118-1122.

30. Livnah, O.; Johnson, D. L.; Stura, E. A.; Farrell, F. X.; Barbone, F. P.; You, Y.; Liu, K. D.; Goldsmith, M. A.; He, W.; Krause, C. D. *Nat. Struct. Biol.* **1998**, *5*, 993-1004.

31. Livnah, O.; Stura, E. A.; Middleton, S. A.; Johnson, D. L.; Jolliffe, L. K.; Wilson, I. A. *Science* **1999**, *283*, 987-990.

32. Syed, R. S.; Reid, S. W.; Li, C.; Cheetham, J. C.; Aoki, K. H.; Liu, B.; Zhan, H.; Osslund, T. D.; Chirino, A. J.; Zhang, J. *Nature* **1998**, *395*, 511-516.

152

33. Moritz, T.; Dutt, P.; Xiao, X.; Carstanjen, D.; Vik, T.; Hanenberg, H.; Williams, D. A. *Blood* **1996**, *88*, 855-862.
34. Hanenberg, H.; Xiao, X. L.; Dilloo, D.; Hashino, K.; Kato, I.; Williams, D. A. *Nat. Med.* **1996**, *2*, 876-882.
35. Miyoshi, H.; Smith, K. A.; Mosier, D. E.; Verma, I. M.; Torbett, B. E. *Science* **1999**, *283*, 682-686.
36. Rollins, C. T.; Rivera, V. M.; Woolfson, D. N.; Keenan, T.; Hatada, M.; Adams, S. E.; Andrade, L. J.; Yaeger, D.; van Schravendijk, M. R.; Holt, D. A. *Proc. Natl. Acad. Sci. USA* **2000**, *97*, 7096-7101.
37. Danielian, P. S.; White, R.; Hoare, S. A.; Fawell, S. E.; Parker, M. G. *Mol. Endocrinol.* **1993**, *7*, 232-240.
38. Suzuki, C.; Ueda, H.; Mahoney, W. C.; Nagamune, T. *Anal. Biochem.* **2000**, *286*, 238-246.

Chapter 12

Beneficial Effects of Silkworm Hemolymph on the Cultivation of Insect Cell-Baculovirus System

Tai Hyun Park, Won Jong Rhee, and Eun Jeong Kim

School of Chemical Engineering, Institute of Chemical Processes,
Seoul National University, Seoul 151–744, Korea

Silkworm hemolymph presents various beneficial effects on the cultivation of insect cell-baculovirus system. It can be used as a substitute for fetal bovine serum which is a medium supplement. It increases the recombinant protein expression rate and host cell longevity after baculoviral infection. The increase in the host cell longevity is due to its apoptosis-inhibition activity. There are several components in the silkworm hemolymph which have apoptosis-inhibition activity. The component with highest activity is a non-glycosylated monomeric protein with the molecular weight of ca. 28,000 Da.

The insect cell-baculovirus system has wide applicability as an alternative to classical bacterial or yeast systems used for the production of recombinant proteins. This system has several advantages, including high expression owing to a strong polyhedrin promoter, production of functionally and immunogenetically active recombinant proteins due to proper post-translational modifications, and nonpathogenicity of the baculoviruses to vertebrates and plants. The key issues of the insect cell-baculovirus system have been concerned with the high cell density culture technology on a large-scale (1-5), the development of an efficient process for the production of recombinant proteins or baculoviruses (6-10),

© 2002 American Chemical Society

media development including low serum and serum-free media (*11, 12*), and the development of high-expression host cell/vector systems (*13, 14*). Recombinant gene expression begins at approximately 1 day post-infection and continues until the host cells die. Host cell viability decreases with time post-infection during production of the recombinant protein. Host cell viability is important for replication of the baculovirus DNA containing a recombinant gene and expression of the cloned gene. Silkworm hemolymph presents beneficial effects on the production of recombinant protein and the longevity of host cell in the insect cell-baculovirus system. Silkworm hemolymph is the most studied insect hemolymph and can be collected easily and economically since the silkworm is a large and cheap domesticated insect. In this chapter, beneficial effects of silkworm hemolymph will be described for the cultivation of insect cell-baculovirus system.

Medium Supplement

Fetal bovine serum (FBS) is most widely used as a medium supplement in insect cell cultures. However, FBS presents some problems including high cost, nonreproducibility due to lot-to-lot variation, undefined composition, increased contamination risk from mycoplasma, and the complication of downstream processing due to a high protein concentration (*15*). Many attempts have been made to develop serum-free medium and there have been several reports of insect cell cultures in serum-free media. However, in most cases, the medium has been supplemented with 10% FBS. In the early stages of insect tissue culture development, silkworm hemolymph was used as a culture medium. Wyatt *et al.* chemically analyzed the silkworm hemolymph (*16*) and formulated a synthetic medium (*17*), which was fortified by Grace (*18*). The Grace's medium still had to be supplemented with insect hemolymph. Ever since FBS was proven to be beneficial for the growth of insect cell, insect cell medium has been supplemented with FBS instead of insect hemolymph. FBS has been used as a supplement since it contains a large number of different growth-promoting activities in a physiologically balanced blend.

Ha *et al.* reported that by adding 5% silkworm hemolymph, FBS concentration in the medium could be reduced to 1% without decreasing the cell growth rate and maximum cell concentration (*19*). Silkworm hemolymph darkens visibly during the incubation due to the activity of tyrosinase in hemolymph, producing melanin via intermediary quinones (*20*). The production of toxic quinones in the medium consequently inhibits the insect cell growth. Therefore, several tyrosinase inhibitors and heat treatment have been suggested for the inactivation of tyrosinase. Heat teatment at 60°C for 30 min is enough for the inactivation of tyrosinase. Cells need to be adapted to grow in reduced FBS medium supplemented with silkworm hemolymph through a gradual adaptation process (*21*).

Production of Recombinant Protein

In the typical Sf9 cell culture, total cell number, which is the sum of viable and dead cell numbers, is maintained at a constant level after baculovirus infection, whereas viable cell number begins to decrease exponentially 3 days after infection. In the case of β-galactosidase production, intracellular product is detected beginning 1 day after infection and reaches a maximum 3 days after infection, while the β-galactosidase begins to be released into a medium 3 days after infection. To investigate the effect of silkworm hemolymph, cells need to be adapted to grow in the medium supplemented with silkworm hemolymph (*19*), and are infected with the recombinant baculovirus when the cell density reaches a late exponential phase.

Silkworm hemolymph when added to medium increases the production of recombinant β-galactosidase in Sf9-AcNPV system (*22*). In the medium supplemented with 1% silkworm hemolymph, intracellular β-galactosidase reaches a maximum concentration and begins to be released into the medium 4 days after infection. The maximum value of the intracellular concentration is about six times as high as that in the medium without hemolymph. Total concentration, that is the sum of intracellular and extracellular concentrations, increases by three-fold with the addition of 1% silkworm hemolymph. In the medium supplemented with 3% or 5% silkworm hemolymph, the production behavior is almost same as that in the medium supplemented with 1% silkworm hemolymph until 4 or 5 days after infection. Total concentration reaches a plateau 4 or 5 days after infection in 1% and 3% silkworm hemolymph, while it keeps on increasing afterwards in 5% silkworm hemolymph. In 5% silkworm hemolymph, final value of total concentration is 4.5 times as high as that without silkworm hemolymph. The addition of more silkworm hemolymph does not increase the production of β-galactosidase; the production profiles in the medium containing 10% silkworm hemolymph shows similar behavior to that with 5% silkworm hemolymph.

The final β-galactosidase total activity and expression per unit cell on the 9th day after infection are summarized in Table I (data from reference *22*). The expression of the cloned gene product increases drastically by adding even 1% silkworm hemolymph. Silkworm hemolymph not only increases the expression rate of the cloned gene but also increases the period of production. Silkworm hemolymph affects also the location of the cloned gene product. For the higher concentration of silkworm hemolymph, more cloned gene product remains inside the cell (*23*).

156

Table I. Effect of Silkworm Hemolymph on Production of β-Galactosidase

Silkworm Hemolymph (%)	Final Concentration of β-Galactosidase (10^3 unit/mL)	β-Galactosidase Expression per Unit Cell (10^{-3} unit/cell)
0	10.2	17.0
1	30.3	59.4
3	30.9	66.3
5	46.5	81.9

Source: Data from Reference 22

As mentioned earlier, FBS concentration in a medium can be reduced to 1% without a decrease in the growth rate and maximum concentration of insect cells by adding 5% silkworm hemolymph. However, the cloned-gene expression is lower in the lower FBS concentration. The lower concentration of FBS gives the lower expression of β-galactosidase in the medium containing 5% silkworm hemolymph; however, it is noteworthy that by adding 5% silkworm hemolymph, the expression in the medium containing only 1% FBS is higher than that in the silkworm hemolymph-free medium containing 5% FBS.

Host Cell Viability

For the efficient production of cloned-gene protein in an insect cell-baculovirus system, not only the high rate of cloned-gene expression but also the viability of host cell is important factor. For higher production of recombinant protein, the host cells need to remain viable longer, and silkworm hemolymph increases the longevity of host cell.

As mentioned earlier, total cell density is maintained at a constant level after baculovirus infection, whereas viable cell density begins to decrease exponentially 3 days after the infection. The baculovirus-induced insect cell death process can be divided into two characteristic phases: a delay phase and a first-order death phase (24, 25). After the baculovirus infection, the host cell viability defined by the ratio of viable cells to total cells is constantly maintained in the delay phase which is characterized by a delay time (t_d). In the first-order death phase which follows the delay phase, cell viability decreases exponentially. This phase is characterized by a specific death rate (k_d) which is a rate constant in a first-order kinetic equation.

The addition of silkworm hemolymph increases the longevity of host cell after baculovirus infection. The delay times and specific death rates at various concentrations of silkworm hemolymph are listed in Table II (23). The delay time increases and the specific death rate decreases as the silkworm hemolymph concentration is higher. Viability is constantly maintained at high level for about 7 days after infection in the medium supplemented with 10% silkworm hemolymph, while host cells begin to die 3 days after infection in the medium without hemolymph. The delay time increases from 3 days to 4.9 days by adding even 1% silkworm hemolymph. The specific death rate (k_d) decreases gradually from 13.8×10^{-3} h^{-1} to 6.0×10^{-3} h^{-1} with the concentration of hemolymph in the medium. By adding 10% silkworm hemolymph, the specific death rate was reduced to 1/2.3 of that without silkworm hemolymph, while the delay time increased 2.3-fold. The lower concentration of FBS gives the lower expression of cloned gene as mentioned earlier; however, the host cell longevity is almost same regardless of FBS concentration. The delay time does not decrease at lower concentration of FBS, whereas lower concentration of silkworm hemolymph results in shorter delay time.

Table II shows that t_d x k_d is nearly constant for every concentration of silkworm hemolymph. This result can be interpreted by the n-target inactivation model which was originally derived to explain the survival rate of cells upon irradiation (26-28) and applied to the baculovirus-induced insect cell death (24). A more detailed mathematical model for the mechanistic steps in infection such as attachment, internalization, endosomal sorting, endosomal fusion, and nuclear accumulation was also developed (29). The number of inactivation targets "n" can be determined by extrapolating the straight line in the first-order death phase to the viability axis. The original meaning of the n was the number of inactivation targets on the cell; however, 'extrapolation number' has been proposed to be used as a term for the n since the model is highly simplified one (26).

The extrapolation number is considered as a measure of the virus-host interaction in the baculovirus-insect cell system (24). According to the simplified n-target inactivation model, $k_d t_d$ remains constant if n is constant. Therefore, the result that k_d x t_d remains almost same in every concentration of silkworm hemolymph in Table II, means that n remains constant regardless of silkworm hemolymph concentration. This indicates the silkworm hemolymph does not affect the number of hypothetical targets representing a host cell susceptibility to the baculovirus.

Inhibition of Apoptosis

The baculovirus-induced insect cell death is known to be apoptosis (30), and cells undergoing apoptosis activate an endonuclease that cleaves DNA between nucleosomes to give fragments. This DNA digestion is the important feature of

Table II. Effect of Silkworm Hemolymph on Delay Time and Specific Death Rate

Silkworm Hemolymph (%)	Delay Time (h)	Specific Death Rate $(10^{-3}h^{-1})$
0	72	13.8
1	117	8.5
3	125	8.0
5	149	6.7
10	164	6.0

apoptosis. The DNA fragmentation does not occur during the first 3 days, while it occurs afterwards in the medium without silkworm hemolymph. Whereas the DNA fragmentation does not occur until the 6th day in the medium supplemented with 5% silkworm hemolymph. As mentioned earlier, the host insect cell viability remains at a high level for 6 days after baculovirus infection in the medium supplemented with 5% silkworm hemolymph, whereas its viability begins to decrease 3 days after infection in a medium without silkworm hemolymph. These DNA fragmentation and cell viability results significantly represent that the time when the DNA fragmentation begins corresponds to the time when the cell viability starts to decrease (23). The inhibition of apoptosis by silkworm hemolymph can be also confirmed by the TUNEL assay (31). Figure 1 shows the typical result of TUNEL assay. In this assay, DNA strand breaks are identified by fluorescence, and the more fluorescence represents the more apoptosis. The baculovirus-infected cells cultured in a medium without silkworm hemolymph shows more fluorescence than those treated with silkworm hemolymph. The flow cytometric analysis shows the apoptosis inhibiting effect of silkworm hemolymph more quantitatively (32). All these results represent that the silkworm hemolymph inhibits the baculovirus-induced insect cell apoptosis.

The addition of silkworm hemolymph either before or during the infection is effective for the inhibition of apoptosis; however, addition after the infection was not so effective. This difference may be explained by a higher transfer rate of silkworm hemolymph into the cells during the virus internalization step. There are two distinct routes by which a virus may enter a cell: endocytosis or fusion. A baculovirus enters a cell largely by endocytosis, although fusion at the plasma membrane cannot be ruled out (33). This endocytosis enhances the transfer of silkworm hemolymph into a cell, consequently, the addition of silkworm hemolymph either before or during the infection will be more effective than after the infection.

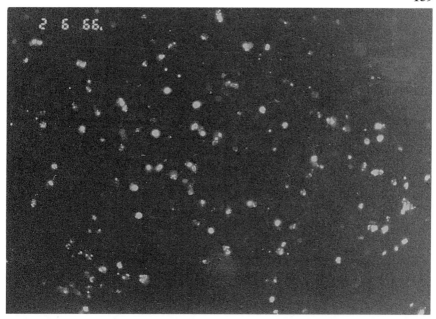

Figure 1. Inhibition effect of silkworm hemolymph on DNA strand breaks. Cells were either not treated (a) or treated (b) with silkworm hemolymph.

A baculovirus first attaches to the host cell surface and then enters the cell. Most viruses attach to cells through a highly specific binding between surface proteins on the virus and cell surface receptor molecules on the host cell membrane (33), although the insect cell receptor that mediates a baculovirus attachment and infection has not yet been identified. Nonetheless, the delayed cell death due to silkworm hemolymph is not caused by the inhibition of these attachment and internalization steps, as silkworm hemolymph is still effective even when it is added after the infection step. As mentioned earlier, the number of hypothetical targets in the n-target inactivation model, considered as a measure of the host cell susceptibility to a baculovirus, is not affected by silkworm hemolymph. This is further evidence that silkworm hemolymph does not inhibit the infection.

The mechanism of baculovirus-induced insect cell apoptosis is not fully understood; however, it is surmised that the apoptosis in insect cells involves a cascade of caspase activation and Sf-caspase-1 is the principal effector caspase in Sf9 cells. In the insect cell-baculovirus system, silkworm hemolymph may work directly on the baculovirus-induced apoptosis cascade mechanism or increase the expression of the anti-apoptotic baculoviral gene such as p35. Recently, it was found that silkworm hemolymph inhibits not only the baculovirus-induced apoptosis but also the apoptosis induced by various chemicals such as actinomycin D, camptothecin, and staurosporine. These results indicate that silkworm hemolymph itself contains anti-apoptotic components itself.

Components with an apoptosis inhibiting activity can be purified from the silkworm hemolymph by heat treatment, gel filtration chromatography, and ion exchange chromatography. Characterization by PAS staining, isoelectric focusing, MALDI-TOF mass spectrometry, and N-terminal sequencing revealed that the component with highest activity was a non-glycosylated monomeric protein with the molecular weight of ca. 28,000 Da (34).

Acknowledgments

The authors wish to acknowledge the financial support of the Korea Science & Engineering Foundation through the Nano Bio-Electronic & System Center.

References

1.   Maiorella, B.; Inlow, D.; Shauger, A.; Harano, D. Bio/technol. 1988, 6, 1406-1410.

2. Murhammer, D. W.; Goochee, C. F. *Bio/technol.* **1988**, *6*, 1411-1418.
3. Reuveny, S.; Kim, Y. J.; Kemp, C. W.; Shiloach, J. *Biotechnol. Bioeng.* **1993**, *42*, 235-239.
4. Lee, S. H.; Park, T. H. *Biotechnol. Lett.* **1994**, *16*, 327-332.
5. Kim, J. H.; Park, T. H. *Biotchnol. Techniques* **1999**, *13*, 425-429.
6. Lindsay, D. A.; Betenbaugh, M. J. *Biotechnol. Bioeng.* **1992**, *39*, 614-618.
7. Wang, M-Y.; Kwong, S.; Bentley, W. E. *Biotechnol. Prog.* **1993**, *9*, 355-361.
8. Lee, S. H.; Park, T. H. *Biotechnol. Techniques* **1995**, *9*, 719-724.
9. Kim, J. H.; Kim E. J.; Park, T. H. *Bioprocess Eng.* **2000**, *23*, 367-370.
10. Inumaru, S.; Kokuho, T.; Yada, T.; Kiuchi, M.; Miyazawa, M. *Biotechnol. Bioprocess Eng.* **2000**, *5*, 146-149.
11. Wilkie, G. E. I.; Stockdale, H.; Pirt, S. V. *Develop. Biol. Stand.* **1980**, *46*, 29-37.
12. Miltenburger, H. G. In Hormonally Defined Media: A Tool in Cell Biology; Fisher, G.; Wieser, R. J. Eds.; Springer-Verlag: New York, NY, 1982; pp 31-43.
13. Wickham, T. J.; Davis, T.; Granados, R. R.; Shuler, M. L.; Wood, H. A. *Biotechnol. Prog.* **1992**, *8*, 391-396.
14. Licari, P.; Jarvis, D. L.; Bailey, J. E. *Biotechnol. Prog.* **1993**, *9*, 146-152.
15. Zhang, J.; Kalogerakis, N.; Behie, L.; Iatrou, K. *Biotechnol. Bioeng.* **1992**, *40*, 1165-1172.
16. Wyatt, G. R.; Loughheed, T. C.; Wyatt, S. S. *J. Gen. Physiol.* **1956**, *39*, 853-868.
17. Wyatt, S. S. *J. Gen. Physiol.* **1956**, *39*, 841-852.
18. Grace, T. D. C. *Nature* **1962**, *195*, 788-789.
19. Ha, S. H.; Park, T. H.; Kim, S.-E. *Biotechnol. Techniques* **1996**, *10*, 401-406.
20. Kim, E. J.; Choi, J.-Y.; Kim, S-E.; Park, T. H. *Biotechnol. Bioprocess Eng.* **1998**, *3*, 87-90.
21. Kim, E. J.; Park, T. H. *J. Microbiol. Biotechnol.* **1999**, *9*, 227-229.
22. Ha, S. H.; Park, T. H. *Biotechnol. Lett.* **1997**, *19*, 1087-1091.
23. Rhee, W. J.; Kim, E. J.; Park, T. H. *Biotechnol. Prog.* **1999**, *15*, 1028-1032.
24. Wu, S.-C.; Dale, B. E.; Liao, J. C. *Biotechnol. Bioeng.* **1993**, *41*, 104-110.
25. Wu, S.-C.; Jarvis, D. L.; Dale, B. E.; Liao, J. C. *Biotechnol. Prog.* **1994**, *10*, 55-59.
26. Alper, T.; Gillies, N. E.; Elkind, M. M. *Nature* **1960**, *186*, 1062-1063.
27. Atwood, K. C.; Norman, A. *Proc. Natl. Acad. Sci. U.S.A.* **1949**, *35*, 696-709.
28. Condon, E. U.; Terrill, H. M. *J. Cancer Res.* **1951**, *11*, 324-333.
29. Dee, K. U.; Shuler, M. L. *Biotechnol. Bioeng.* **1997**, *54*, 468-490.
30. Clem, R. J.; Fechheimer, M.; Miller, L. K. *Science* **1991**, *254*, 1388-1390.

162

31. Rhee, W. J.; Park, T. H. *Biochem. Biophys. Res. Commun.* **2000**, *271*, 186-190.
32. Rhee, W.J.; Park, T.H. *J. Microbiol. Biotechnol.* **2001**, *11*, 853-857
33. Hammer, D. A.; Wickham, T. J.; Shuler, M. L.; Wood, H. A.; Granados, R. R. In Baculovirus Expression Systems and Biopesticides; Shuler, M. L.; Wood, H. A.; Granados, R. R.; Hammer, D. A. Eds.; Wiley-Liss, Inc.: New York, NY, 1995; pp 103-119.
34. Kim, E. J.; Rhee, W. J.; Park, T. H. *Biochem. Biophys. Res. Commun.* **2001**, *258*, 224-228.

Chapter 13

Apoptotic Signal Transduction by Cadmium Ion and Detoxification by Plant Peptides

Masahiro Takagi[1], Hiroyuki Satofuka[1], and Tadayuki Imanaka[2]

[1]School of Materials Science, Japan Advanced Institute of Science and Technology, 1–1 Asahidai, Tatsunokuchi, Ishikawa 923–1292, Japan
[2]Department of Synthetic and Biological Chemistry, Graduate School of Engineering, Kyoto University, Sakyo, Kyoto 606–8501 Japan

Apoptotic cell death of Jurkat cells by Cd^{2+} toxicity was studied by fluorescence microscopic observation and DNA fragmentation assay. It was suggested that apoptotic response by Cd^{2+} was less clear than that by a typical apoptosis inducer, ultraviolet (UV, 254nm). Examination of MAP kinase phosphorylation (p38, JNKs and c-Jun) by Cd^{2+} toxicity indicated that the phosphorylation was very slowly activated (>4 h after stimulation), while UV could activate the phosphorylation immediately. Therefore, it was suggested that Cd^{2+} may not be a typical apoptosis inducer. Antioxidants (glutathione (GSH) and N-acetylcysteine (NAC)) could detoxify Cd^{2+} cooperatively with Bcl-2. By addition of a plant-specific peptide, phytochelatin (PC$_7$, (γGlu-Cys)$_7$-Gly) to the media, detoxification of Cd^{2+} and the cooperation with Bcl-2 were more intensive than the cases of GSH and NAC. PC synthase gene was transferred from *Arabidopsis thaliana* to the Jurkat cell. The transformant exhibited resistance to Cd^{2+} and production of plant-specific PC (PC$_{2-6}$).

© 2002 American Chemical Society

163

164

INTRODUCTION

Cd^{2+} is an environmental pollutant with well-known mutagenic, carcinogenic and teratogenic effects (1). Cd^{2+} is known to accumulate in the human kidney for a relatively long time and at high doses, and is also known to produce health effects on the respiratory system and has been associated with bone diseases (1, 2). At molecular and cellular levels, a lot of studies about apoptosis and stress kinase activation by Cd^{2+} have been performed (3-7). However, possibly because diverse cellular responses by toxicity of Cd^{2+} were included, mechanism of the toxicity is not fully understood.

Plants respond to heavy metal toxicity via a number of mechanisms. One such mechanism involves the chelation of heavy metal ions by a family of peptide ligands, the phytochelatins (PCs, $(\gamma Glu\text{-}Cys)_n\text{-}Gly$, $n \geq 2$) (8-10). Recently we reported rapid method for detection and detoxification of heavy metal ions in water environments using PCs (11). Moreover, genes encoding PC synthase from *Arabidopsis thaliana*, *Schizosaccharomyces pombe* and wheat were reported. These genes, designated *AtPCS1 (CAD1)*, *SpPCS* and *TaPCS1*, respectively, encode 40-50% sequence-similar 50-55 kDa polypeptides active in the synthesis of PC from GSH (12-14). Mammalian cells can not synthesize PC because of their deficiency in the key enzyme for phytochelatin biosynthesis, PC synthase.

In this study, we examined apoptotic cell death caused by Cd^{2+} and activation of stress kinases in comparison with a typical apoptosis inducer, UV (254 nm). Furthermore, we attempted to express the plant-specific peptides (PCs) in mammalian cells by transforming Jurkat cells with a plant PC synthase gene (*AtPCS1*).

MATERIALS AND METHODS

Antibodies. Antibodies against p38, JNKs, c-Jun and their phosphorylated forms were purchased from New England Biolabs (Beverly, MA, USA). Alkaline phosphatase-conjugated secondary antibody was from ICN Biochemicals Inc. (Costa Mesa, CA, USA).

Phytochelatin. Homogeneous phytochelatin (PC_7) was chemically synthesized by slightly modified Fmoc-method as explained in our previous work (11).

Cell treatments and cell viability assay. Jurkat cells, the human leukemic T cell line, were used for examination, and the cells transfected with pUC-CAGGS vector (15) inserted with human *bcl-2* gene (16) (Jurkat Bcl-2 cells) and with pUC-CAGGS vector (Jurkat CAGGS cells) were prepared. All cells were cultured in RPMI 1640 medium supplemented with 10% fetal bovine serum (FBS) at 37 °C and 5% CO_2. Exponentially growing cells at 60-80% confluency were diluted to a concentration of 5×10^4 cells/ml, then a stock solution of 10 mM $CdCl_2$ was added to the culture medium at final concentrations of 10-100 µM. Cells were incubated for 48 h. For UV irradiation, cells were placed in a tissue culture dish and were exposed to UV at an intensity of 1 W/m^2 for 1 min (about 40 J/m^2) by the UV-C (254 nm) light source (Spectroline ENF-260 C/J, Spectronics Co., NY, USA). Cell viability was measured using Cell Titer 96 AQ_{ueous} Cell Proliferation assay kit (Promega, Madison, WI, USA).

Observation of apoptosis and DNA fragmentation. Fluorescence microscopic assay was performed as described previously (17) to determine the apoptotic cell death. Hoechst 33342 and propidium iodide were added to the medium to final concentration of 30 µM and 10 µM, respectively. After incubation at 37 °C for 10 min, cells were examined using a fluorescence microscope (Olympus BX50, Tokyo, Japan) with UV excitation at 360 nm. Nuclei of viable, necrotic, early apoptotic and late apoptotic cells were observed as blue round, pink round, fragmented blue and fragmented pink, respectively. DNA fragmentation was detected by ApopLadder EX kit (Takara Biomedicals, Kyoto, Japan).

Western blotting. Whole cell proteins were fractionated on 12% SDS-PAGE and transferred to polyvinylidene difluoride membrane (Millipore, Bedford, MA, USA). After blocking with 5% non-fat milk in TBS-T buffer (30 mM Tris-HCl, 150 mM NaCl, 0.1% (v/v) Tween 20, pH 7.5), the membrane was incubated with primary antibody at 4 °C for 16 h, followed with extensive wash with TBS-T buffer. Then the membrane was incubated with secondary antibody at room temperature for 1 h. After the washing step, the immune complexes were detected by enhanced chemiluminescence (ECL) system (Amersham Pharmacia Biotech Uppsala, Sweden).

Plasmid construction. The cDNA for PC synthase gene (*AtPCS1*) was amplified by polymerase chain reaction (PCR) using the cDNA phage library of *A. thaliana* (Arabidopsis Biological Resource Center, Columbus, OH, USA) as a template. For the PCR reaction, the primers (PSF1, 5'-ACCATGGCTATGG CGAGTTTATATCGGCG-3' (containing the *Nco*I site indicated by underline); PSR1, 5'-CGGGATCCCTAATAGGCAGGAGCAGCGAGATCATCC-3' (carrying the *Bam*HI site as underlined)) were used for amplification of 1.5 kbp cDNA containing whole coding region of the PC synthase. The amplified cDNA was digested with *Bam*HI, and subcloned into pUC18 plasmid digested with *Hin*cII and *Bam*HI (Takara Biomedicals, Kyoto, Japan). The cDNA was digested with *Hin*dIII and *Bam*HI, and the fragment was inserted into pcDNA3 vector (Invitrogen, Carlsbad, CA, USA) and the resultant plasmid was designated as pcDNA3-PCS.

Detection of PC synthesis in Jurkat cells. The postcolumn derivatization method using DTNB (5,5'-dithiobis-2-nitrobenzoic acid) was carried out for sensitive detection of PCs in extract of Jurkat cells. Cells were exposed to 20 µM $CdCl_2$ for 40 h, and harvested by centrifugation ($100 \times g$, 5 min). After addition of 700 µl of 0.5 N NaOH containing 0.5 mg/ml $NaBH_4$, cells were sonicated for 3 min and kept on ice for 10 min to reduce the disulfides. The extracted sample were neutralized by addition of 1 ml of 3.6 N HCl to stop the reducing reaction by $NaBH_4$ and centrifuged ($5000 \times g$, 10 min, 4 °C) to remove the insoluble debris. The clear supernatant was subject to reversed-phase high-performance liquid chromatography (HPLC) (column: Hibar HPLC-cartrige 250-4 LiChrospher 100RP-18 (5 µM); solvent system: A: 0.02% TFA, 5.0 mM octanesulfonic acid, B: 100% acetonitrile, 0.02% TFA, 5.0 mM octanesulfonic acid; gradient 13–100% B in 30 min; flow rate 1.5 ml/min). And the eluted sample was continuously mixed with the thiol-reactive solution (10% acetonitrile, 75 µM DTNB, 50 mM potassium phosphate, pH 8.0) to detect the PC and GSH at 412 nm.

RESULTS

Observation of apoptotic bodies and DNA fragmentation. Cells were exposed to $CdCl_2$ and apoptotic responses such as morphological changes in cell nucleus and fragmentation of chromosomal structure into nucleosomal units (ca. 180 bp) were examined. Results were compared with the case of UV irradiation (254nm), a typical apoptotic stress signal. Ratio of apoptotic bodies among the total dead cells by Cd^{2+} toxicity (ca. 4%) was much lower than that by UV irradiation (ca. 70%) (data not shown). Moreover, smear DNA fragmentation could be observed only around 30 µM of $CdCl_2$ (Figure 1, lane 3) and no DNA fragmentation could be detected at higher concentrations (50 µM) (Figure 1, lane 4), although very clear DNA fragmentation could be observed in the case of UV (Figure 1, lane 1). These results indicated that Cd^{2+} could not induce typical apoptotic cell death.

Activation of p38 and JNKs signal pathways by Cd^{2+}. Activation of MAP kinase pathways by Cd^{2+} toxicity was studied by detecting phosphorylated p38, JNKs and c-Jun in the cells exposed to $CdCl_2$. At 30 µM of $CdCl_2$, phosphorylation of p38 and JNKs in Jurkat CAGGS were induced at 4 h and 8 h after the exposure, respectively (Figure 2B). It was also shown that stimulation

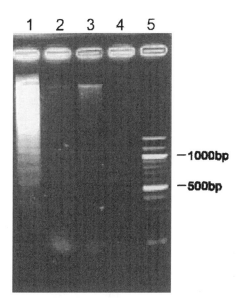

Figure 1: DNA fragmentation by Cd^{2+}. Jurkat cells $(5 \times 10^4$ cells/ml) were exposed to UV for 1 min (lane 1), control (lane 2), 30 μM $CdCl_2$ (lane 3) or 50 μM $CdCl_2$ (lane 4) for 6 h, respectively. Lane 5 is for molecular size markers.

168

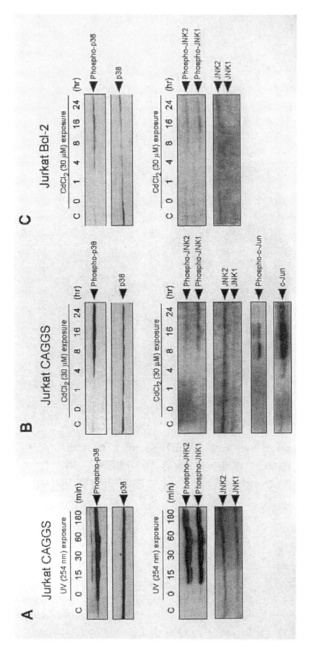

Figure 2: Phosphorylation of p38, JNKs and c-Jun in response to treatments with Cd^{2+} and UV. The figures shows the relative levels of phosphorylated p38 and JNKs in Jurkat CAGGS cells treated with UV (A), treated with 30 μM CdCl$_2$ (B) and in Jurkat Bcl-2 cells treated with 30 μM CdCl$_2$ (C). The phosphorylation of c-Jun in Jurkat CAGGS cells treated with 30 μM CdCl$_2$ was also shown (B). Lane C represents the control with untreated cell extract. The proteins of p38, JNKs and c-Jun, and those phosphorylated forms were detected by corresponding antibodies. Two bands with molecular weight of 46 kDa and 55 kDa detected by anti-JNKs antibody were corresponding to JNK1 (37) and JNK2 (38), respectively.

of p38 pathway by Cd^{2+} was stronger than that of JNKs pathway (Figure 2B). In the case of UV, levels of phosphorylated p38 and JNKs were markedly increased soon after the irradiation (Figure 2A). These results indicated that p38 and JNKs pathways were very slowly activated by Cd^{2+}, while the pathways were immediately activated by UV. This delayed and weak activation of MAP kinases is consistent with the above-mentioned results of microscopic observation and DNA fragmentation assay of apoptotic cell death by Cd^{2+}.

Effects of Bcl-2 on Cd^{2+} toxicity. Bcl-2 is a well-known suppressor of apoptosis (18, 19). When Jurkat Bcl-2 cells expressing higher level of Bcl-2 were exposed to Cd^{2+}, the intensity of expression and phosphorylation of p38 and JNKs were weaker than the case of Jurkat CAGGS cells (Figure 2, B and C). Therefore, to investigate whether Bcl-2 can suppress Cd^{2+} toxicity in Jurkat cells, Jurkat Bcl-2 cells were exposed to 10-100 μM $CdCl_2$ and cell viability was tested. Suppression of Cd^{2+} toxicity by Bcl-2 could be observed at 20-60 μM $CdCl_2$ (Figure 3A), but most intensive effect could be observed at 30 μM. Jurkat Bcl-2 cells were completely resistant up to 30 μM $CdCl_2$, while 40% of the control cells could not be survived at 30 μM $CdCl_2$.

Effects of antioxidants on Cd^{2+} toxicity. Jurkat Bcl-2 and Jurkat CAGGS cells were incubated respectively in complete media containing respective antioxidants, GSH and NAC (equivalent to 500 μM thiol group) with 10-100 μM $CdCl_2$. By addition of GSH or NAC to the medium, suppression of Cd^{2+} toxicity could be observed within the range of 30-70 μM $CdCl_2$, and effect of NAC was slightly stronger than that of GSH (Figure 3B). When Jurkat Bcl-2 cells were used, cell growth was completely inhibited at 70 μM of $CdCl_2$. However, GSH and NAC cooperatively suppressed Cd^{2+} toxicity with Bcl-2 at wider range of $CdCl_2$ concentrations (Figure 3C).

Detoxification of Cd^{2+} by PC. Phytochelatin (PC) is heavy metal ion binding peptide produced by plant, algae and some fungi (8-10). By addition of the chemically synthesized PC_7 (equivalent to 500 μM thiol group) to the culture media containing 10-100 μM $CdCl_2$, cell survival ratio of Jurkat CAGGS cells was clearly recovered (Figure 4A). Furthermore, in the case of Jurkat Bcl-2 cells, the cooperative suppression of PC_7 and Bcl-2 against Cd^{2+} toxicity became more significant (Figure 4B) than the cases of GSH and NAC (Figure 3C). Interestingly cell survival ratio of Jurkat Bcl-2 cells in the medium containing PC_7 exceeded 100% in the range of 40-60 μM of $CdCl_2$.

Transformation of Jurkat cells with PC synthase gene. To investigate whether PC can be synthesized in mammalian cells by introduction of the PC synthase gene (*AtPCS1*) from *A. thaliana*, the *AtPCS1* gene was amplified by

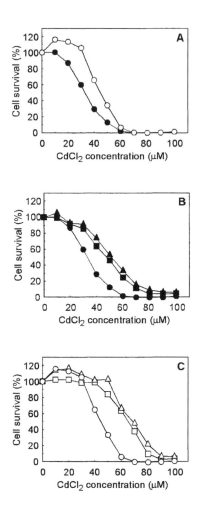

Figure 3: Effect of Bcl-2 on Cd^{2+} toxicity and cooperative detoxification with antioxidants (GSH or NAC). Jurkat Bcl-2 (open circles) and Jurkat CAGGS cells (closed circles) were treated with 10-100 μM CdCl$_2$ (A). Cell survival ratio in the media containing CdCl$_2$ (10-100 μM) with GSH (squares), NAC (triangles) or without antioxidant (circles) was assayed for both Jurkat CAGGS (B) and Jurkat Bcl-2 cells (C). Antioxidants, GSH and NAC equivalent to 500 μM of thiol group were added to the medium, respectively.

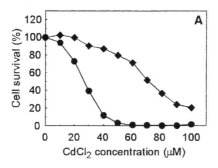

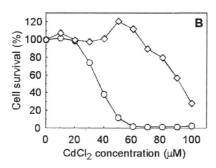

Figure 4: Cooperative effect of PC$_7$ and Bcl-2 against Cd^{2+} toxicity. The cell survival ratio of Jurkat CAGGS cells (A) and Jurkat Bcl-2 (B) treated with 10-100 µM CdCl$_2$ with (diamonds) or without PC$_7$ (circles). PC$_7$ equivalent to 500 µM of thiol group was added to the medium.

PCR and inserted into pcDNA3 vector to construct pcDNA3-PCS. The plasmid DNA was used to transform Jurkat cells and transformants were selected by an antibiotic (G418). Transformation of Jurkat cells with pcDNA3-PCS was confirmed by PCR and fluorescence *in situ* hybridization (data not shown). The constructed transformant, Jurkat cells carrying *AtPCS1*, was designated as Jurkat PCS. Tolerance to $CdCl_2$ (10-50 μM) and PC synthesis were examined, and compared with the case of control cells transformed with pcDNA3 (Jurkat DNA3 cells). Survival ratio of Jurkat PCS cells was significantly higher than that of Jurkat DNA3 cells (Figure 5), indicating that PC synthase were expressed in mammalian cells and functioned properly.

Cell lysate of the Jurkat PCS cells was analyzed by HPLC with postcolumn derivatization method as explained above. In the case of Jurkat PCS cells treated with 20 μM $CdCl_2$, PC synthesis corresponding to PC_2-PC_6 could be clearly observed (Figure 6A) indicating that PC synthase gene from *A. thaliana* could catalyze PC synthesis even in Jurkat cells. It was reported that activity of PC synthase from *A. thaliana* was posttranslationally enhanced by Cd^{2+} (20). Indeed, activity of the PC synthase in Jurkat cells was also stimulated with Cd^{2+}. Therefore, not only catalytic activity but also activation mechanism of the plant PC synthase was functional in the mammalian cells.

DISCUSSION

It has been reported that Cd^{2+} could induce apoptosis in various types of cells (21-26). In this study, however, we showed that Cd^{2+}-induced apoptotic cell death was not in dose-dependent manner, and both apoptotic bodies and DNA fragmentation were not clear in comparison with the typical apoptosis induced by UV irradiation (Figure 1). Western blotting analyses clearly showed that stress kinases such as p38 and JNKs were very slowly activated (> 4 h) (Figure 2), although they were activated immediately after the stimuli of UV irradiation. These observations indicated that Cd^{2+} was not necessarily a strong inducer of apoptosis. The death signal by Cd^{2+} toxicity must be diversified to different signal cascade(s) causing cell death.

At present, by observation of the suppression effect of Bcl-2 or antioxidants (NAC or GSH) toward Cd^{2+} toxicity (Figure 3, A and B), we hypothesized that Cd^{2+}-induced stress is a kind of oxidative stress. Indeed, toxicity of the Cd^{2+} has been reported to interfere with antioxidant enzymes and radical scavengers such as superoxide dismutases, peroxidases and catalases (27, 28). Higher level of cytosolic reactive oxygen species (ROS), known as a cell death signals (29), produced by Cd^{2+} might be removed by these enzymes and scavengers. Bcl-2 exerts broad anti-apoptotic effects by inhibiting the production of ROS and enhancing the steady state of mitochondrial transmembrane functions (18). In our experimental results, MAP kinase activation in case of Jurkat Bcl-2 was weaker than the case of control cells although the interval before activation was the same (> 4 h after stimulation) (Figure 2C). Cd^{2+}-induced cell death could be very effectively suppressed by cooperative effect between antioxidants and Bcl-2 (Figure 3C). These results further supported that Cd^{2+} toxicity is diversified including a kind of oxidative stress and relating to ROS production.

Effect of detoxification by PC added to the medium was much stronger than that by other antioxidants (Figure 4A). However, in terms of cooperation, PC and Bcl-2 was more significant than the cases of GSH and NAC (Figure 4B), indicating that PC might be functional for not only physically chelating of Cd^{2+} but also suppressing diversified toxicity of Cd^{2+} including oxidative stress and ROS production.

PC deficient plants are reported to be very sensitive to Cd^{2+} (30). We assumed that these PCs are important antioxidants to decrease harmful oxidative stresses of UV and heat as well as heavy metal ions. It is interesting to examine how plant-specific PCs can function in mammalian cells whose vacuoles are not well developed. Cell extracts of the transformant, Jurkat PCS cells, apparently contained PCs with different sizes (PC_{2-6}) (Figure 6A). Interestingly, the length of the PCs found in transformants of mammalian cells is longer than those reported for plant cells (PC_{2-4}) (31). Without any physical stresses, intracellular concentration of GSH in mammalian cells is generally kept at 1-8 mM (32). On the contrary, in *A. thaliana* cells except chloroplast, the concentration of GSH is maintained at lower concentrations (50-250 μM) (33). This difference in intracellular GSH concentration might be the reason for synthesis of longer PCs in transformed mammalian cells. Nevertheless, it is interesting to note that the plant *AtPCS1* gene can be utilized to enhance resistance to Cd^{2+} toxicity of mammalian cells (Figure 5).

One possible mechanism of diverse death signals might be initiated from membrane lipid peroxidation by ROS induced by Cd^{2+}. The process of cell death caused by Cd^{2+} toxicity might be closely related to inflammatory process *via* arachidonic cascade (34, 35) rather than apoptotic signal cascade represented by stress kinase activation. Recently it was reported that p38 is more important for oxidative stress and inflammatory process (36) than JNK pathways. Our observation about stress kinase activation also showed that activation of p38 signal was slightly stronger than that of JNKs signals, although both signals were equally, strongly and immediately activated by UV (Figure 2A). Therefore, we considered that this stronger activation of p38 might be characteristic for death signal started from ROS, and the result further supported important linkage between oxidative stress and inflammatory process.

Further studies about ROS production, membrane lipid peroxidation, intracellular calcium ion concentration and NO synthesis by toxicity of Cd^{2+} are now in progress.

174

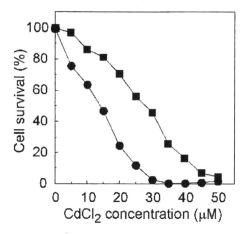

Figure 5: Cd^{2+} tolerance of *AtPCS1* gene expressing Jurkat cells. Jurkat PCS (carrying pcDNA3-PCS plasmid) (squares) and Jurkat DNA3 cells (carrying the empty pcDNA3 plasmid) (circles) were treated with 10-50 μM CdCl$_2$ for 48 h.

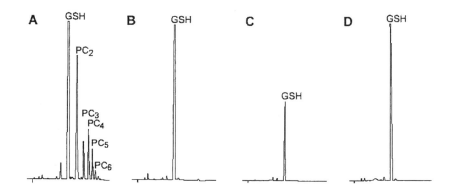

Figure 6: PC synthesis in Jurkat cells mediated by *AtPCS1* gene. Cell extracts of Jurkat PCS and Jurkat DNA3 cells were analyzed by HPLC with postcolumn derivatization method. Jurkat PCS cells were treated with 20 μM CdCl$_2$ (A) and left untreated (B). As an experimental control, Jurkat DNA3 cells were also treated with (C) or without (D) 20 μM CdCl$_2$. Peaks designated "GSH", "PC$_2$", "PC$_3$", "PC$_4$", "PC$_5$" and "PC$_6$" were identified as such on the basis of their retention times, respectively, compared with native PCs extracted from *Silene cucubalus* (31).

175

REFERENCES

1. Waalkes, M. P. *J. Inorg. Biochem.* **2000**, *79*, 241-244.
2. Berglund, M.; Akesson, A.; Bjellerup, P.; Vahter, M. *Toxicol. Lett.* **2000**, *112*, 219-225.
3. Matsuoka, M.; Igisu, H. *Biochem. Biophys. Res. Commun.* **1998**, *251*, 527-532.
4. Iordanov, M. S.; Magun, B. E. *J. Biol. Chem.* **1999**, *274*, 25801-25806.
5. Chuang, S. M.; Wang, I. C.; Yang, J. L. *Carcinogenesis* **2000**, *21*, 1423-1432.
6. Ding, W.; Templeton, D. M. *Toxicol. Appl. Pharmacol.* **2000**, *162*, 93-99.
7. Galan, A.; Garcia-Bermejo, M. L.; Troyano, A.; Vilaboa, N. E.; de Blas, E.; Kazanietz, M. G.; Aller, P. *J. Biol. Chem.* **2000**, *275*, 11418-11424.
8. Zenk, M. H. *Gene* **1996**, *179*, 21-30.
9. Rauser, W. E. *Plant Physiol.* **1995**, *109*, 1141-1149.
10. Cobbett, C. S. *Plant Physiol.* **2000**, *123*, 825-832.
11. Satofuka, H.; Amano, S.; Atomi, H.; Takagi, M.; Hirata, K.; Miyamoto, K.; Imanaka, T. *J. Biosci. Bioeng.* **1999**, *88*, 287-292.
12. Clemens, S.; Kim, E. J.; Neumann, D.; Schroeder, J. I. *EMBO J.* **1999**, *18*, 3325-3333.
13. Ha, S. B.; Smith, A. P.; Howden, R.; Dietrich, W. M.; Bugg, S.; O'Connell, M. J.; Goldsbrough, P. B.; Cobbett, C. S. *Plant Cell* **1999**, *11*, 1153-1164.
14. Vatamaniuk, O. K.; Mari, S.; Lu, Y. P.; Rea, P. A. *Proc. Natl. Acad. Sci. USA* **1999**, *96*, 7110-7115.
15. Niwa, H.; Yamamura, K.; Miyazaki, J. *Gene* **1991**, *108*, 193-199.
16. Shimizu, S.; Eguchi, Y.; Kamiike, W.; Matsuda, H.; Tsujimoto, Y. *Oncogene* **1996**, *12*, 2251-2257.
17. Shimizu, S.; Eguchi, Y.; Kamiike, W.; Itoh, Y.; Hasegawa, J.; Yamabe, K.; Otsuki, Y.; Matsuda, H.; Tsujimoto, Y. *Cancer Res.* **1996**, *56*, 2161-2166.
18. Voehringer, D. W.; Meyn, R. E. *Antioxid. Redox Signal.* **2000**, *2*, 537-550.
19. Tsujimoto, Y.; Shimizu, S. *FEBS Lett.* **2000**, *466*, 6-10.
20. Vatamaniuk, O. K.; Mari, S.; Lu, Y. P.; Rea, P. A. *J. Biol. Chem.* **2000**, *275*, 31451-31459.
21. Habeebu, S. S.; Liu, J.; Klaassen, C. D. *Toxicol. Appl. Pharmacol.* **1998**, *149*, 203-209.
22. Hart, B. A.; Lee, C. H.; Shukla, G. S.; Shukla, A.; Osier, M.; Eneman, J. D.; Chiu, J. F. *Toxicology* **1999**, *133*, 43-58.
23. Achanzar, W. E.; Achanzar, K. B.; Lewis, J. G.; Webber, M. M.; Waalkes, M. P. *Toxicol. Appl. Pharmacol.* **2000**, *164*, 291-300.
24. Fujimaki, H.; Ishido, M.; Nohara, K. *Toxicol. Lett.* **2000**, *115*, 99-105.
25. Kim, M. S.; Kim, B. J.; Woo, H. N.; Kim, K. W.; Kim, K. B.; Kim, I. K.; Jung, Y. K. *Toxicology* **2000**, *145*, 27-37.

26. Szuster-Ciesielska, A.; Stachura, A.; Slotwinska, M.; Kaminska, T.; Sniezko, R.; Paduch, R.; Abramczyk, D.; Filar, J.; Kandefer-Szerszen, M. *Toxicology* **2000**, *145*, 159-171.
27. Hussain, T.; Shukla, G. S.; Chandra, S. V. *Pharmacol. Toxicol.* **1987**, *60*, 355-358.
28. Muller, L. *Toxicology* **1986**, *40*, 285-295.
29. Irani, K. *Circ. Res.* **2000**, *87*, 179-183.
30. Inoue, M.; Ito, R.; Ito, S.; Sasada, N.; Tohoyama, H.; Joho, M. *Plant Physiol.* **2000**, *123*, 1029-1036.
31. Gill, E.; Winnacker, E. L.; Zenk, M. H. *Methods in Enzymology* **1991**, *205*, 333-341.
32. Griffith, O. W. *Free Radic. Biol. Med.* **1999**, *27*, 922-935.
33. Gutierrez-Alcala, G.; Gotor, C.; Meyer, A. J.; Fricker, M.; Vega, J. M.; Romero, L. C. *Proc. Natl. Acad. Sci. USA* **2000**, *97*, 11108-11113.
34. Miyahara, T.; Tonoyama, H.; Watanabe, M.; Okajima, A.; Miyajima, S.; Sakuma, T.; Nemoto, N.; Takayama, K. *Calcif. Tissue Int.* **2001**, *68*,185-191.
35. Romare, A.; Lundholm, C. E. *Arch. Toxicol.* **1999**, *73*, 223-228.
36. Bellmann, K.; Burkart, V.; Bruckhoff, J.; Kolb, H.; Landry, J. *J. Biol. Chem.* **2000**, *275*, 18172-18179.
37. Derijard, B.; Hibi, M.; Wu, I. H.; Barrett, T.; Su, B.; Deng, T.; Karin, M.; Davis, R. *J. Cell* **1994**, *76*, 1025-1037.
38. Su, B.; Jacinto, E.; Hibi, M.; Kallunki, T.; Karin, M.; Ben-Neriah, Y. *Cell* **1994**, *77*, 727-736.

Chapter 14

Chemotherapy with Hybrid Liposomes Composed of Dimyristoylphosphatidylcholine and Polyoxyethylenealkyl Ether without Drugs

Ryuichi Ueoka[1], Yoko Matsumoto[2], Hideaki Ichihara[2], and Tetsushi Kiyokawa[3]

Departments of [1]Applied Life Science and [2]Applied Chemistry, Sojo University, Ikeda, Kumamoto 860–0082, Japan
[3]Kumamoto National Hospital, Ninomaru, Kumamoto 860–0008, Japan

We produced hybrid liposomes composed of dimyristoyl-phosphatidylcholine and polyoxyethlenealkyl ether having the uniform and stable structure. Highly inhibitory effects of hybrid liposomes were obtained on the growth of tumor cells in vitro. Induction of apoptosis by hybrid liposomes through activation of caspase-3 was verified using flow cytometry and microphysiometer. Significantly prolonged survival was obtained in mice inoculated tumor cells after the treatment with hybrid liposomes without any side effect. In clinical applications, prolonged survival was attained in patients with lymphoma after the intravenous injection of hybrid liposomes without drugs and the remarkable reduction of solid tumor was obtained after the local administration of hybrid liposomes.

It is well known that liposomes are used as drug carriers : example are antitumor agents, hormones, and immunomodulation (1). In particular, polyoxyethyleneglycol (PEG) – phosphatidylcholine (PC) liposomes have been found to be effective for prolonging blood circulation (2). On the other

© 2002 American Chemical Society

178

hand, we have recently produced specific hybrid liposomes composed of vesicular and micellar molecules ; they are free from any contamination from organic solvents and stable for a longer period. Changing the composition of hybrid liposomes can control the physical properties of these liposomes, such as size, membrane fluidity, phase transition temperature, and hydrophobicity (3,4). A schematic representation of hybrid liposomes is shown in Fig. 1.

In the course of our study on the liposomes, the following interesting results were obtained. (a) Stereochemical control of the enantioselective hydrolysis of amino acid esters could be established by temperature regulation and by changing the composition of hybrid liposomes (3,4,5,6,7). (b) Inhibitory effects of hybrid liposomes including flavonoids (8) or antitumor drugs (9, 10) on the growth of tumor cells in vitro and in vivo have been obtained. (c) Remarkably high inhibitory effects of hybrid liposomes on the growth of tumor cells in vitro have been obtained without drugs (11, 12, 13, 14). (d) No toxicity of the hybrid liposomes was observed in normal cells in vitro and in normal rats in vivo without any side effects (10, 15).

In this study, we report on the uniform and stable hybrid liposomes composed of dimyristoylphosphatidylcholine (DMPC) and polyoxy-ethylenealkyl ether ($C_{12}(EO)_n$) having inhibitory effects on the growth of tumor cells in vitro, in vivo and clinical chemotherapy.

$$CH_3(CH_2)_{12}COCH_2$$
$$CH_3(CH_2)_{12}COCH$$
$$O \quad CH_2O - P - OCH_2CH_2N(CH_3)_3$$
$$O^-$$

DMPC

$$CH_3(CH_2)_{11}O(CH_2CH_2O)_{23}H$$

$C_{12}(EO)_{23}$

In Vitro Antitumor Effects

We examined inhibitory effects of hybrid liposomes composed of DMPC and $C_{12}(EO)_n$ on the growth of human promyelocytic leukemia (HL-60) cells in vitro. The cells were cultured for 4 days in a 5% CO_2 incubator at 37.C after adding the hybrid liposomes. The inhibitory effects of hybrid liposomes on the growth of tumor cells were evaluated by 100(Na-Np)/Na, when Na and Np denote the live cell numbers in the absence and presence of the hybrid liposomes, respectively. The hybrid liposomes were prepared by dissolving both DMPC and $C_{12}(EO)_n$ in phosphate – buffered saline with sonication (BRANSONIC MODEL B2210 apparatus, 90W) at 45.C for 5 min. The results are summarized in Table Ⅰ. The noteworthy aspects are as follows : (a) The inhibitory effects of the hybrid liposomes composed of DMPC and $C_{12}(EO)_4$ are moderately enhanced as compared with those of the liposomes of DMPC. (b) The inhibitory effects of DMPC/$C_{12}(EO)_8$ and DMPC/$C_{12}(EO)_{23}$ hybrid liposomes were fairly enhanced. (c) Almost

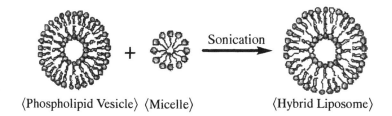

⟨Phospholipid Vesicle⟩ ⟨Micelle⟩   ⟨Hybrid Liposome⟩

Figure 1. Schematic representation of hybrid liposomes.

Table 1. Inhibitory Effects of Hybrid Liposomes.

liposomes	inhibitory effect, %	
DMPC	25	
DMPC/10mol% $C_{12}(EO)_4$	64	(0)
DMPC/10mol% $C_{12}(EO)_8$	89	(77)
DMPC/10mol% $C_{12}(EO)_{10}$	96	(83)
DMPC/10mol% $C_{12}(EO)_{12}$	99	(75)
DMPC/10mol% $C_{12}(EO)_{23}$	79	(3)

$[DMPC]=7.5 \times 10^{-5}M$, $[C_{12}(EO)_n]=8.3 \times 10^{-6}M$
Initial cell number : 1.0×10^4 cells/ml
Values in the parentheses are those of $C_{12}(EO)_n$ micells.
The inhibitory effects have maximum error of $\pm 6\%$

180

completely inhibitory effects were attained by employing the hybrid liposomes of DMPC/$C_{12}(EO)_{10}$ and DMPC/$C_{12}(EO)_{12}$. These results suggest that the hydrophilic-hydrophobic balance in polyoxyethylenealkyl ether is important for the enhancement of the inhibitory effects on the growth of tumor cells.

We also examined the morphology of the hybrid liposomes composed of 90mol% DMPC and 10mol% ($C_{12}(EO)_n$) on the basis of electron microscopy and dynamic light scattering measurements. The electron micrograph of the hybrid liposomes shows the presence of spherical vesicles (Fig. 2). Interestingly, a clear stock solution of the hybrid liposomes having a hydrodynamic diameter of 70 nm with a single and narrow distributions could be kept over 30 days, as shown in Fig. 3. These results indicate that the uniform and stable structure of hybrid liposomes composed of 90mol% DMPC and 10mol% $C_{12}(EO)_{10}$ could be obtained.

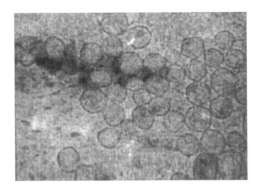

Figure 2. Electron micrograph of hybrid liposomes composed of 90mol% DMPC and 10mol% $C_{12}(EO)_{12}$.

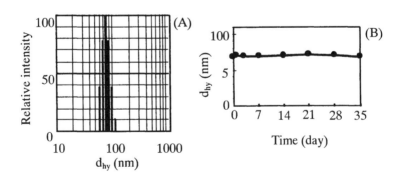

Figure 3. Distribution of d_{hy} (A) and time course of d_{hy} change (B) for hybrid liposomes composed of DMPC and $C_{12}(EO)_{10}$.

Antitumor Mechanism

The induction of apoptosis for HL-60 cells by the hybrid liposomes composed of DMPC and $C_{12}(EO)_{10}$ were examined using flow cytometry and microphysiometer. The DNA content was determined in a flow cytometer with propidium iodide staining method. The results are shown in Fig. 4. It is noteworthy that exposure of HL-60 cells to the hybrid liposomes of DMPC/$C_{12}(EO)_{10}$ caused fragmented DNA characteristic of apoptosis (16).

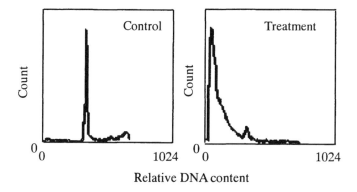

Relative DNA content

Figure 4. Rerative DNA contents for HL-60 cells treated with the hybrid liposomes composed of 90mol% DMPC and 10mol% $C_{12}(EO)_{10}$.

We examined intracellular responses of the hybrid liposomes composed of DMPC and $C_{12}(EO)_{10}$ against HL-60 cells related to apoptosis. Intracellular responses of HL-60 cells after the addition of hybrid liposomes were measured by the Cytosenser microphysiometer (Molecular Devices). The Cytosenser microphysiometer measures the change in extracellular acidification rate resulting either from alterations in the energy demand made on the cells as they respond to the effector agents or from alterations in sodium-hydrogen exchange across the cell membrane (17). The results are shown in Fig. 5. A rapid and transient increase in the acidification rate was observed immediately after adding hybrid liposomes to the culture medium. This transient increase in acidification is probably due to a massive release to protons from inside to outside of cells following the opening of ion channels. After the initial response, the acidification rate gradually increased, to reach a maximum value, and thereafter it slowly decreased. It is presumed that the increase in acidification was caused by activation of some membrane proteins in the HL-60 cells after their fusion with hybrid liposomes, and the decrease in acidification was caused by subsequent apoptosis of HL-60 cells (18). It was suggested that the almost constant value of acidification rate for 6 h might reach to the completion of apoptosis. Relative DNA contents and the time course of accumulation of DNA fragmented by DMPC/10mol%$C_{12}(EO)_{10}$ hybrid liposomes (in the same concentration as in Fig. 5) are shown in Fig. 6. The DNA of HL-60 cells was found to be almost completely fragmented 6 h after the addition of hybrid liposomes, indicating the completion of apoptosis. It is noteworthy that the fairly agreement of completion time for the apoptosis between microphysiometry and flow cytometry experiments was observed.

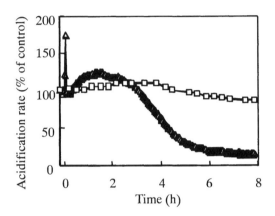

Figure 5. Apoptotic DNA rate for HL-60 cells treated with hybrid liposomes composed of 90mol% DMPC and 10mol% $C_{12}(EO)_{10}$.
(□:control, Δ: DMPC/10mol%$C_{12}(EO)_{10}$)
[DMPC] = 5.0×10^{-4}M, [$C_{12}(EO)_{10}$] = 5.5×10^{-5}M

Concentration dependences of caspase-3, 7 and caspase-3 inhibitor on the apoptotic DNA rate were obtained as shown in Fig. 7. These results suggest that the hybrid liposomes should induce apoptosis through the activation of caspase-3.

We represent a hypothetic mechanism for the inhibitory effects of hybrid liposomes on the growth of tumor cells as shown in Fig. 8. Hybrid liposomes should fuse to tumor cell membrane and activate caspase. As a result, fragmentation of DNA should be occurred.

In Vivo Antitumor Effects

We examined the inhibitory effects of hybrid liposomes in mice intraperitoneally inoculated with B16 melanoma cells in vivo (*19, 20*).

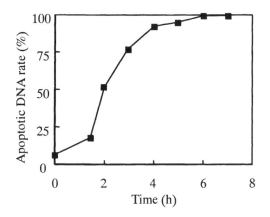

Figure 6. Apoptotic DNA rate for HL-60 cells treated with hybrid liposomes composed of 90mol% DMPC and 10mol% $C_{12}(EO)_{10}$.
[DMPC] = 5.0×10^{-4}M, [$C_{12}(EO)_{10}$] = 5.5×10^{-5}M

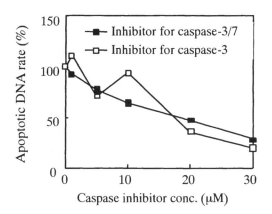

Figure 7. Inhibition of apoptotic DNA rate for HL-60 cells treated with hybrid liposomes composed of 90mol% DMPC and 10mol% $C_{12}(EO)_{10}$ in the presence of inhibitor toward caspase-3/-7 or caspase-3.

184

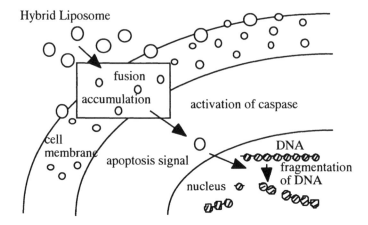

Figure 8. A hypothetic mechanism for the inhibitory effects of hybrid liposomes on the growth of tumor cells.

These cells were originally from skin tumors but can be taken as a model of lung cancer, since melanoma frequently metastasizes to the lymph nodes and lungs. Screening for therapeutic effects is generally carried out in mice or rats bearing tumor cells. In the primary screening, tumor cells are introduced into the peritoneal cavity, and the test agent is administered intraperitoneally. Animals were randomly grouped on the basis of body weight on the day of tumor cell inoculation using the stratified randomization method. B16 melanoma cells (5×10^5 cells) were inoculated intraperitoneally. Hybrid liposomes were administered 13 times starting from 1h after the inoculation every day. The median life span was calculated using the equation : (median survival days af ter treatment)/(median survival days of control group)×100. The results are shown in Fig. 9. The median life span was 103% and 115% in the $C_{12}(EO)_{10}$-treated group and $C_{12}(EO)_{23}$-treated group, respectively. It is noteworthy that significantly prolonged survival (162-170%) was obtained in mice treated with both the hybrid liposomes of DMPC/10mol%$C_{12}(EO)_{10}$ and DMPC/10mol%$C_{12}(EO)_{23}$.

Furthermore, a mouse model of lung carcinoma was established by intraperitoneal inoculation of Lewis lung carcinoma cells (*21*). The survival times of mice model of lung carcinoma in the control groups were 25.0 and 45.0 days for median and maximum values, respectively, as shown in Fig. 10. It is noteworthy that the survival times in the treated groups with hybrid liposomes of DMPC/10mol%$C_{12}(EO)_{10}$ were more than one year. These results suggest that hybrid liposomes should be effective for the treatment of lung carcinoma. It is also of interest that the survival mice should be perfectly cured after the treatment of hyb rid liposomes over one year.

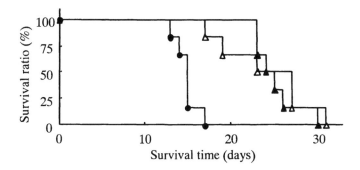

Figure 9. Survival curves of mice treated with hybrid liposomes after the intraperitoneal inoculation of B16 melanoma cells. 6 Mice were employed in each experiment. Dose : DMPC, 680mg/kg. ●,control; Δ, DMPC/10mol% $C_{12}(EO)_{10}$; ▲,DMPC/10mol% $C_{12}(EO)_{23}$

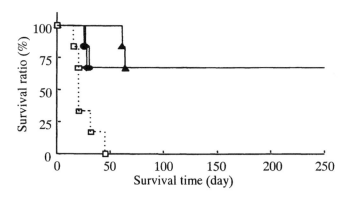

Figure 10. Survival curves of mice treated with or without hybrid liposomes composed of 90mol% DMPC and 10mol%$C_{12}(EO)_{10}$ after the intraperitoneal inoculation of Lewis lung carcinoma cells. □,control; ♦,Treatment 5ml/kg; ●, Treatment 10ml/kg; ▲,Treatment 20ml/kg

186

Clinical Application

Patients with lymphoma were selected for the clinical application of hybrid liposomes composed of DMPC and $C_{12}(EO)_{23}$ without any drug after passing the committee of bioethics. The clinical diagnosis for one patient was as follows: (a) No effects of all chemotherapies were observed.   (b) It was suggested that the future lifetime would be a few months.   So, we tried to examine the treatment with hybrid liposomes for that patient.   The hybrid liposomes were administered intravenously every day. There were not abnormal findings on routine blood test (fig. 11) and hematochemistry (Fig. 12).

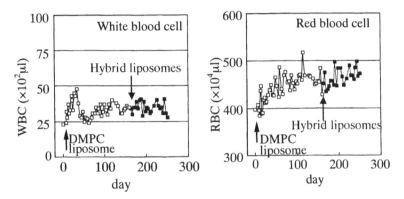

Figure 11. Findings on routine blood test after the intravenous injection of the hybrid liposomes composed of DMPC and $C_{12}(EO)_{23}$.

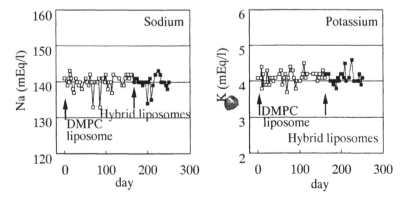

Figure 12. Findings on routine hemato chemistry after the intravenous injection of the hybrid liposomes composed of DMPC and $C_{12}(EO)_{23}$.

The high γ-GTP values returned to normal ones after the administration of the hybrid liposomes having a diameter of 100nm, though the administration of DMPC liposomes caused an increase in γ-GTP (Fig. 13). This result indicates that the hybrid liposomes could avoid the reticules endothelial system (RES). Prolonged survival was attained in patient with lymphoma after the intravenous injection of hybrid liposomes without any side effect and the remarkable reduction of neoplasm (solid tumor) was obtained after the local administration of hybrid liposomes (Fig. 14).

Such a successful clinical application of hybrid liposomes without drugs was for the first time in the world.

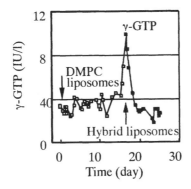

Figure 13. γ-GTP values after the intraveneous administration of DMPC or hybrid liposomes.

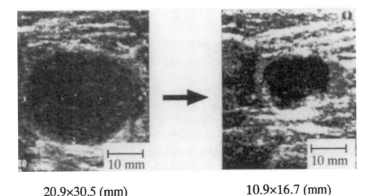

20.9×30.5 (mm)   10.9×16.7 (mm)

Figure 14. Photographs with ultra sonical echo after the treatment of hybrid liposomes.

Conclusion

A summary of the noteworthy aspects of this study is as follows: (a) the uniform and stable structure of hybrid liposomes composed of dimyristoylphosphatidylcholine and polyoxyethlenealkyl et her was obtained. (b) Highly inhibitory effects of hybrid liposomes were obtained on the growth of tumor cells in vitro. (c) Induction of apoptosis by hybrid liposomes through activation of caspase-3 was verified using flow cytometry and microphisiometer. (d) Significantly prolonged survival was obtained in mice inoculated tumor cells after the treatment with hybrid liposomes without any side effect. (e) In clinical application, the prolonged survival was attained in patient with lymphoma after the intravenous injection of hybrid liposomes without drugs and the remarkable reduction of solid tumor was obtained after the local administration of hybrid liposomes.

Acknowledgment

We thank Dr. Akihiro Kanno for his technical assistance in animal experiments and Megumi Yamamoto for her technical assistance in flow cytometry. This work was supported in part by a Grant – in –Aid for Science Research from Ministry of Education, Science, Culture, and Sports of Japan (Nos. 12019270, 12217152, 12555232).

References

1. *Liposomes and Their Uses Biology and Medicine* ; Papahadyopoulos, D., Ed. : Ann. NY Acad. Aci., 1978 ; p308.
2. Allen, T. M. ; Hansen, C. J. ; Martin, F., Redemann, C., Yan, Y. A. *Biochem. Biophys. Acta* **1991**, 1066, 29-36.
3. Ueoka, R. ; Moss, R. A ; Swarup, S. ; Matsumoto, Y. ; Strauss, G. ; Murakami, Y. *J. Am. Chem. Soc.* **1985**, 107, 2185-2186.
4. Ueoka, R. ; Matsumoto, Y. ; Moss, R. A ; Swarup, S. ; Sugii, A. ; Harada, K. ; Kikuchi, J. ; Murakami, Y. *J. Am. Chem. Soc.* **1988,** 110, 1588-1595.
5. Ueoka, R.; Yamada, E,; Yamashita, O.; Matsumoto, Y.; Kato, Y. *Tetrahedron Lett.* **1991,** 32, 6597-6600.
6. Goto, K.; Matsumoto, Y.; Ueoka, R. *J. Org. Chem.* **1995,** 60, 3342-3346.
7. Tanoue, O.; Baba, M.; Tokunaga, Y.; Goto, K.; Matsumoto, Y.; Ueoka, R. *Tetrahedron Lett.* **1999,** 40, 2129-2132.
8. Ueoka, R.; Matsumoto, Y.; Oyama, H.; Takekuma, H.; Iio, M. *Chem. Pharm. Bull.* **1988,** 36, 4640-4643.
9. Matsumoto, Y.; Yamada, E.; Hirano, J.; Oshige, M.; Iio, M.; Iwahara, M.; Ueoka, R. *Biol. Pharm. Bull.* **1993,** 16, 213-215.
10. Kitamura, I.; Kochi, M.; Matsumoto, Y.; Ueoka, R.; Kuratsu, J.; Ushio, Y. *Cancer Res.* **1996,** 56, 3986-3992.

11. Matsumoto, Y.; Imamura, C.; Ito, T.; Taniguchi, C.; Ueoka, R. *Biol. Pharm. Bull.* **1995**, 18, 1456-1458.
12. Imamura, C.; Kemura, Y.; Matsumoto, Y.; Ueoka, R. *Biol. Pharm. Bull.* **1997**, 20, 1119-1121.
13. Matsumoto, Y.; Kato, T.; Iseki, S.; Suzuki, H.; Nakano, K.; Iwahara, M.; Ueoka, R. *Bioorg. Med. Chem. Lett.* **1999**, 9, 2617-2619.
14. Matsumoto, Y.; Kato, T.; Suzuki, H.; Hirose, S.; Naiki, Y.; Hirashima, M.; Ueoka, R. *Bioorg. Med. Chem. Lett.* **2000**, 10, 1937-1940.
15. Imamura, C.; Kanno, A.; Mitsuoka, C.; Kitajima, S.; Inoue, H,; Iwahara, M.; Matsumoto, Y.; Ueoka, R. *Yakugaku Zasshi,* **1996**, 116, 942-947.
16. Matsumoto, Y.; Kato, T.; Kemura, Y.; Tsuchiya, M.; Yamamoto, M.; Ueoka, R. *Chem. Lett.* **1999**, 53-54.
17. Parce, J.W.; Owicki, J.C.; Kercso, K.M.; Signal, G.B.; Wada, H.G.; Muir, V.C. Bousse, L.J.; Ross, K.L. Sikic, B.I.; McConnell, H.M. *Science,* **1989**, 246, 243-247.
18. Kuo, R.C.; MacEwan, D.J.; Boxster, G.T. *Mol. Biol. Cell.,* **1994**, 5, 26-27.
19. Kanno, A.; Tsuzaki, Y.; Miyagi, M.; Matsumoto, Y.; Ueoka, R. *Biol. Pharm. Bull.* **1999**, 22, 1013-1014.
20. Kanno, A.; Kodama, R.;Terada, Y.; Matsumoto, Y.; Ueoka, R. *Drug Delivery System,* **1998**, 13, 101-105.
21. Kanno, A.; Terada, Y.; Tsuzaki, Y.; Matsumoto, Y.; Ueoka, R. *Drug Delivery System,* **1999**, 14, 37-42.

Chapter 15

Apoptosis Inhibiting Genes and Caspase Inhibitors Improved Mammalian Cell Survival and Enhanced Protein Production

Satoshi Terada[1], Akiko Ogawa[1], Naoto Sakai[1], Masao Miki[1],
Tetsuo Fujita[2], Tsuyoshi Yata[2], Teruyuki Nagamune[2],
Norio Sakuragawa[3], and Eiji Suzuki[4]

[1]Department of Applied Chemistry and Biotechnology, Faculty of
Engineering, Fukui University, 3–9–1, Bunkyo, Fukui 910–8507, Japan
[2]Department of Chemistry and Biotechnology, The University of Tokyo,
7–3–1, Hongo, Bunkyo-ku, Tokyo 113–8656, Japan
[3]Department of Inherited Metabolic Disorders, National Institute of
Neuroscience, 4–1–1, Ogawa-Higashi, Kodaira, Tokyo 187–8502, Japan
[4]Research Institute of Innovative Technology for the Earth, 9–2,
Kizugawadai, Kizu-cho, Soraku-gun, Kyoto 619–0292, Japan

Inhibiting apoptosis would prolong culture period and could
contribute increasing protein productivity per culture. In this
study, over-expression of apoptosis inhibiting genes and
supplementation of caspase inhibitors were investigated.
Overexpression of bcl-2 in hybridoma cells prolonged the
culture and successfully increased antibody production.
Addition of caspase-3 inhibitor, tetra-peptide DEVD, to the
culture of bcl-2 transfectant was effective for delaying cell
death but failed to increase antibody production.
Overexpression of bcl-2 delayed cell death of CHO cells in
serum deprived culture, while transfection of caspase
inhibiting genes crmA from cowpox virus or p35 from
baculovirus did slightly. Caspase inhibitor delayed cell death

© 2002 American Chemical Society

of wild type CHO cells but did not synergically delayed the bcl-2 transfectant. Caspase inhibitors delayed cell death of human amniotic epithelial cells. Caspase-3 inhibitor successfully prolonged the in vitro culture period of human amniotic epithelial cells and EGF increased this effect.

Introduction

Useful protein production of a culture would increase when the viable culture period extends (Fig.1). However, hybridoma and CHO cells, producer of useful protein, tend to die quickly after reaching the maximum cell density. Therefore, preventing cells from death which starts in the late exponential growth phase and maintaining them viable in batch culture for longer time period should increase protein production of the culture.

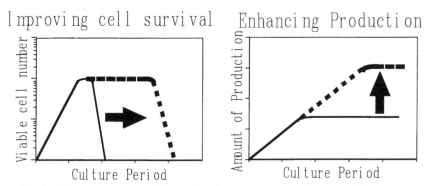

Fig.1 Schematic strategy illustrating how increase production of bio-active products synthesized by mammalian cells in batch culture.

Mammalian cells in culture are exposed to environmental changes, including accumulation of toxic metabolites, consumption of nutrients and growth factors, pH, DO and so on. At least one of these harmful conditions occurs in the late exponential growth and stationary phases of batch culture and triggers cell death. Apoptosis is described as suicidal death and Bcl-2 protein has been found to be functional in suppressing apoptosis (1), (2). As shown in Fig.2, apoptosis (or programmed cell death) signal pathway is complicated. Caspase-mediated pathway or caspase-independent, and blocked by Bcl-2 or not. Synergistical inhibition, therefore, would be desirable for delaying cell death which starts in the late exponential growth phase, because at the over-growth phase, multiple adverse conditions such as nutrient or serum component depletion and toxic metabolite accumulation induce apoptosis through different pathways.

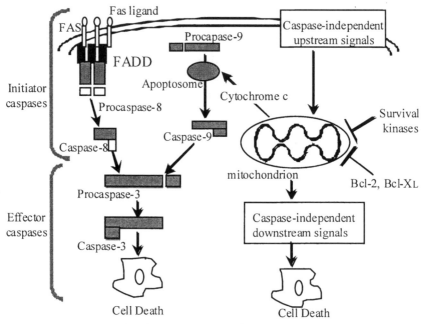

Fig.2 Apoptosis signaling pathway
Caspases are activated in hierarchical order. First initiator proteases,
including caspase-6, 8, and 9, are activated. Then the initiator caspases activate
effector caspases and result in cell death. Death trigers upstream of
mitochondrion usually do not require activation of caspases. Chytochrome c
induces caspase activation and triggers typical apoptosis.

ICE/CED-3 family proteases such as caspase-1 and caspase-3 play a key role in apoptosis (3). These proteases are constitutively expressed in cells as inactive precursors and they require cleavage into two subunits about 20 and 10 Kd and tetramerization before activation. After activated, they cleave several cellular proteins and the cleavage appears to be an early event of apoptosis as it occurs before any morphological signs of cell death (4), (5). Two viral proteins, CrmA from cowpoxvirus (6) and p35 and baculovirus (7), are potent inhibitors of the proteases. And synthesized tetrapeptide inhibitors, DEVD as caspase-3 inhibitor and YVAD as caspase-1 inhibitor, were known(8), (9). These tetrapeptide inhibitors were designed with the appropriate peptide recognition sequences of the substrates of each proteases.

In this study, we transfected the cells with anti-apoptotic genes, including bcl-2, crmA and p35, and added tetrapeptide caspase inhibitors to the culture for delaying cell death and for increasing protein production. And in order to increase population of human amniotic epithelial cells, candidate of cell therapy without immune response, tetrapeptide inhibitors were added to the culture.

Materials and Methods

Cell Lines and Culture Condition

A cell line 2E3-O is a mouse hybridoma derived from a mouse myeloma P3X63 AG8U.1 by electric fusion with mouse spleen cells. Chinese hamster ovary (CHO) cells were obtained from Riken Cell Bank (RCB0285, Japan). Human placentas were obtained from uncomplicated elective Caesarean sections in accordance with the requirements of the several hospitals in Kodaira City (Japan) and the human amniotic membranes were mechanically peeled free from the placenta. Amniotic epithelial cells were gotten from the membranes treated with trypsin.

The cells were cultured in DMEM, a-MEM or RPMI 1640 medium (Nissui, Japan), supplemented with 10% FBS (Gibco BRL, USA), 20 mM HEPES, 0.2% NaHCO3, 2 mM glutamine, and 0.06 mg/ml kanamycin. The cells were grown at 37 °C in humidified air containing CO2 at 5%.

Viable and dead cells were determined by counting in a hemocytometer under a phase contrast microscope using trypan blue exclusion.

Transfection

The vector BCMGneo-bcl-2, BCMGneo-p35 BCMGneo-crmA for expressing human bcl-2, baculovirus p35 and cowpox virus crmA, respectively, were prepared. The transfection method was electroporation.

Reagents

As caspase-3 inhibitor, Ac-Asp-Glu-Val-Asp-CHO, DEVD, (Takara, Japan) and as caspase-1 inhibitor, Ac-Tyr-Val-Ala-Asp-CHO, YVAD (Takara, Japan) were used.

194

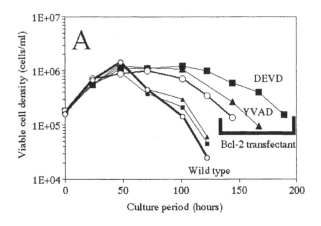

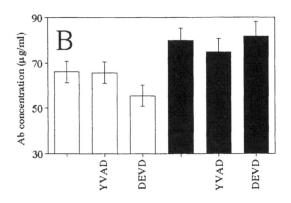

Caspase inhibitor

Fig.3 Effect of introducing bcl-2, apoptosis inhibiting gene, and caspase inhibitor on hybridoma cell proliferation (a), and antibody production (b). Wild type & Bcl-2 transfectant were cultured in DMEM medium supplemented with 10% FBS and with or without caspase inhibitor. (a) Viable cell density was determined by the trypan blue dye exclusion method. Open squares, closed triangles, and closed circles are the result obtained in the culture of wild type in the absence of inhibitors, in the presence of 1 mM caspase-1 inhibitor (YVAD) and in the presence of 1 mM caspase-3 inhibitor (DEVD), respectively. Closed squares, open triangles, and open circles are the result obtained in the culture of Bcl-2 transfectant in the absence of inhibitors, in the presence of 1 mM YVAD and in the presence of 1 mM DEVD, respectively. (b) When culture were terminated, the antibody concentration of the cultures were determined with ELISA. Open bar represents the culture of wild type and closed bars do bcl-2 transfectants.

195

Results and Discussion

Effect of Caspase Inhibitors on Survival of Hybridoma Cell

We tested the effect of 1 mM concentration of casapse-1 inhibitor, YVAD, and caspase-3 inhibitor, DEVD, on the proliferation of hybridoma. The growth curve of the culture is shown in Fig.3-a. After the day 3 of the culture, the viable cell density of the cells treated with DEVD and YVAD were slightly higher than cells untreated. But treatment of caspase inhibitors prolonged the culture less long than bcl-2 transfection.

Effect of Caspase Inhibitors on Survival of Hybridoma Cell Overexpressing Bcl-2

Because multi factors that induce apoptosis exist at the late exponential growth phase and decline phase of the culture, sole inhibitor might not block all of the pathways. Therefore, we jointly applied apoptosis suppressive gene, bcl-2, and tetrapeptide caspase inhibitors, for improving survival of hybridoma cell at over-growth phase.

As shown in Fig.3-a, viable cell density of wild type was below 10,000 cells/ml at day 4 of the culture, that of bcl-2 transfected cells untreated was at day 6, and that of bcl-2 transfected cells treated with caspase-3 inhibitor, DEVD, was at day 8. Jointly application of bcl-2 transfection and addition of DEVD to the culture prolonged the culture period for about four days.

This survival effect also increased the population per unit medium. As shown in Table 1, 1 ml of the medium without inhibitor cultured 2,730,000 wild type hybridoma cells * day, and 3,930,000 bcl-2 transfectant cells * day, while 1 ml of the medium with DEVD cultured 6,080,000 bcl-2 transfectant cells * day.

We determined the antibody (Ab) concentration of culture supernatant by ELISA when the culture was terminated. Ab concentration of culture supernatant was shown in Fig.3-b. The culture of the cells transfected with bcl-2 produced more Ab than the wild type culture. Though the culture period was prolonged amazingly, culture of the bcl-2 transfectants treated with caspase-3 inhibitor failed to increase Ab production.

Caspase inhibitors prolonged the culture period but failed to produce more Ab, which suggests that treatment with caspase inhibitor might suppress the specific Ab production rate per cell per time.

Table 1 Population of hybridoma cells per culture (10^4 cells * day / ml)

	without inhibitor	YVAD	DEVD
Wild type	273	273	328
Bcl-2 transfectant	393	475	608

The cell density in Fig-3-a were integlated with respect to culture time.

Effect of Bcl-2, P35 and CrmA on Survival of CHO Cells

CHO cell line is widely used for recombinant protein production by cell culture. Aiming at more efficient protein production by cell culture, we transfected bcl-2, p35 or crmA gene to CHO cells. The viability of the CHO cells transfected with the gene in serum-deprived culture were shown in Fig. 4-a. At day 2 of the culture, the viability of mock transfectant was 72 %, that of bcl-2 transfectant was 83 % and that of p35 or crmA was 78%. Overexpression of bcl-2 in CHO cells slightly improved survival under serum deprived culture.

Effect of Bcl-2 and Caspase Inhibitor on Survival of CHO Cells

Because the effect of bcl-2 on CHO cell survival was not so well, caspase-3 inhibitor, DEVD, were jointly added to the serum-deprived culture and cell viability was determined at day 7 (Fig.4-b). While the addition of DEVD delayed cell death of mock-transfectant, the addition failed to delay cell death of bcl-2 transfectant. This result may suggest that both bcl-2 and caspase-3 inhibitor block same pathway of cell death.

Bcl-2 transfection delayed the cell death of CHO and this prolonging culture would contribute larger amount of recombinant protein production.

Increase Population of Human Amniotic Epithelial Cells

Because amnion does not induce rejection after allotransplantation, amniotic epithelial cell is candidate for cell therapy (10). But human amniotic epithelial (HAE) cells do not proliferate so much in in vitro culture (11). For increasing population of HAE cells, apoptosis inhibition was tried.

At first, the effect of caspase inhibitors on HAE cell survival was tested. Under serum deprived condition, HAE cell viability was declined and this was recovered by caspase inhibitors. The survival effect of 0.5 mM YVAD and 1 mM YVAD were similar, while 1 mM DEVD delayed cell death as much as 0.5 mM YVAD joined with 0.5 mM DEVD. These results suggest that the pathways blocked by DEVD include not only caspase-1-dependent pathway but also independent pathway.

Long-term culture with DEVD was performed (Fig5-b). At first stage, treatment of DEVD delayed cell death and increased the population, but after 400 hours culture, the increasing effect was not observed. But addition of EGF delayed cell death of the culture with DEVD, while did not without DEVD.

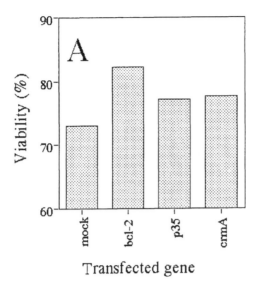

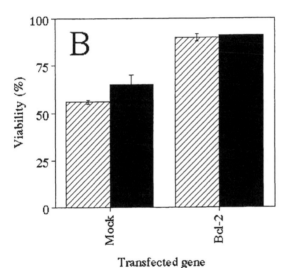

Fig.4 Effect of introducing apoptosis inhibiting genes on CHO cell survival in serum-deprived culture (a) CHO cells transfected with bcl-2, p35, crmA or mock were cultured in serum-free a-MEM for 2 days. Cell viability was determined by the trypan blue dye exclusion method. (b) CHO cells transfected with bcl-2, p35 or mock were cultured in serum-free a-MEM supplemented with DEVD (caspase-3 inhibitor, closed) or without (open). At day 7, cell viability was determined by the trypan blue dye exclusion method.

198

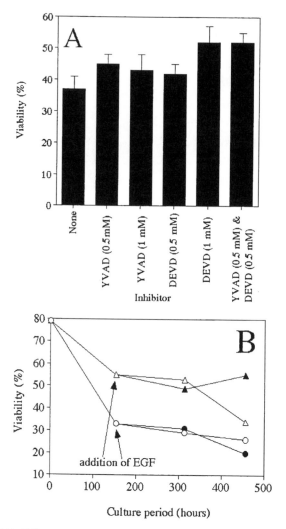

Fig.5 (A) Effect of caspase inhibitor on amniotic epithelial cell survival
under serum deprived culture. Human amniotic epithelial cells were cultured in
serum-free RPMI 1640 medium with caspase inhibitor or without. Cell viability
was determined at day 8. (B) Effect of DEVD and of EGF on amniotic
epithelial cell survival Human amniotic epithelial cells were cultured with 1mM
DEVD (triangle) or without (circle). At day 6, the cultures indicated by the
closed symbols were added with 5 ng/ml of EGF. At day 6 and at day 13, the
medium were replaced with fresh one with or without the reagents. Cell viability
was determined by the trypan blue dye exclusion method.

References

1. Tsujimoto, Y. Stress-resistance conferred by high level of bcl-2 alpha protein in human B lymphoblastoid cell. *Oncogene* **1989**, *4*, 1331-1336
2. Pettersson, M.; Jernberg-Wiklund, H.; Larsson, L.G.; Sundstrom, C.; Givol, I.; Tsujimoto, Y.; Nilsson, K. Expression of the bcl-2 gene in human multiple myeloma cell lines and normal plasma cells. *Blood* **1992**, *79*, 495-502.
3. Martin, S.J.; Green, D.R. Protease activation during apoptosis: death by a thousand cuts? *Cell* **1995**, *82*, 349-352
4. Wilson, K.P.; Black, J.A.; Thomson, J.A.; Kim, E.E.; Griffith, J.P.; Navia, M.A.; Murcko, M.A.; Chambers, S.P.: Aldape, R.A.; Raybuck, S.A. Structure and mechanism of interleukin-1 beta converting enzyme. *Nature* **1994**, *370*, 270-275.
5. Nicholson, D.W.; Ali, A.; Thornberry, N.A.; Vaillancourt, J.P.; Ding, C.K.; Gallant, M.; Gareau, Y.; Griffin, P.R.; Labelle, M.; Lazebnik, Y.A. Identification and inhibition of the ICE/CED-3 protease necessary for mammalian apoptosis. *Nature* **1995**, *376*, 37-43.
6. Miura, M.; Zhu, H.; Rotello, R.; Hartwieg, E.A.; Yuan, J. Induction of apoptosis in fibroblasts by IL-1 beta-converting enzyme, a mammalian homolog of the C. elegans cell death gene ced-3. *Cell* 1993, 75, 653-660.
7. Bump, N.J.; Hackett, M.; Hugunin, M.; Seshagiri, S.; Brady, K.; Chen, P.; Ferenz, C.; Franklin, S.; Ghayur, T.; Li, P. Inhibition of ICE family proteases by baculovirus antiapoptotic protein p35. *Science* **1995**, *269*, 1885-1888.
8. Thornberry, N.A.; Peterson, E.P.; Zhao, J.J.; Howard, A.D.; Griffin, P.R.; Chapman, K.T. Inactivation of interleukin-1 beta converting enzyme by peptide (acyloxy)methyl ketones. *Biochemistry* **1994**, *33*, 3934-3940.
9. Rotonda, J.; Nicholson, D.W.; Fazil, K.M.; Gallant, M.; Gareau, Y.; Labelle, M.; Peterson, E.P.; Rasper, D.M.; Ruel, R.; Vaillancourt, J.P.; Thornberry, N.A.; Becker, J.W. The three-dimensional structure of apopain/CPP32, a key mediator of apoptosis. *Nature Struct Biol.* **1996**, *3*, 619-625.
10 Adinolfi, M.; Akle, C.A.; McColl, I.; Fensom, A.H.; Tansley, L.; Connolly P.; Hsi, B.L.; Faulk, W.P.; Travers, P.; Bodmer, W.F. Expression of HLA antigens, beta 2-microglobulin and enzymes by human amniotic epithelial cells. *Nature*1982, *295*, 325-327
11 Terada, S.; Matsuura, K.; Enosawa, S.; Miki, M.; Hoshika, A.; Suzuki, S.; Sakuragawa, N. Inducing proliferation of human amniotic epithelial (HAE) cells for cell therapy. *Cell transplantation* **2000**, *9*, 701-704

Chapter 16

Presence of Catalytic Activity of the Antibody Light Chain Raised against Complementarity Determining Region Peptide of Super Catalytic Antibody

Y. Zhou, E. Hifumi, H. Kondo, and T. Uda*

Department of Biosciences, Hiroshima Prefectural University, Shobara City, Hiroshima 727-0023, Japan
*Corresponding author: Phone: 81-824-74-1756; fax: 81-824-74-0191; email: uda@bio.hiroshima-pu.ac.jp

A monoclonal antibody i41SL1-2 raised against the peptide of complementarity determining region-1 (CDRL-1) of super catalytic antibody, 41S-2-L, which is capable of emzymatically destoroying the gp41 molecule of HIV-1 envelope, was prepared. The light chain, i41SL1-2-L, catalytically decomposed the CDRL-1 peptide through the successive reaction. Based on the molecular modeling, i41SL1-2 possesses a catalytic triad composed of Ser, His, Asp, whose positions are identical to those of the catalytic antibody, VIPase.

© 2002 American Chemical Society

Introduction

Some interesting natural antibodies possessing high catalytic efficiency have been reported so far (1-5). The characteristic ones are the autoantibodies to vasoactive intestinal polypeptide (VIP) by Paul et al. (1) and to DNA by Gabibov et al. (2). Authors have recently found a novel natural catalytic antibody (referred to as "super catalytic antibody") (6-9). The light chain (41S-2-L) of the antibody could enzymatically decompose the antigenic gp41 peptide (19 mer peptide: a highly conserved region in many HIV-1 strains) as well as the intact gp41 molecule which plays an important role for the entry of human immunodeficiency virus (HIV) into human T cells (10-11). It is revealed that the complementarity determining region-1 (CDRL-1; RSSKSLLYSNGNTYLY) of the 41S-2-L substantially concerns with the antigen recognition (12). In this paper, we will describe that the light chain of the antibody (i41SL1-2) raised against the CDR1 of 41S-2-L has the ability to enzymatically decompose the antigenic CDRL-1 peptide. The molecular modeling suggests the presence of a serine protease-like catalytic site on the surface of the light chain of i41SL1-2.

Experimental

The i41SL1-2 mAb was raised against the synthetic peptide of CDRL-1-Cys (Cys was introduced at the C-terminus in order to conjugate with KLH) by immunizing Balb/c mice. The peptides used in this study were synthesized by the Fmoc solid-phase method using an automated peptide synthesizer (Applied Biosystems 431A, CA, USA). The resultant peptides were purified by reversed-phase HPLC (RP-HPLC; Waters 490E, Waters mBONDASPHERE C_{18} column; Waters, NY, USA) and the purities were confirmed to be over 99% (data not shown). The peptide identification was established using an ion-spray type mass spectrometer (API-III, Perkin-Elmer Sciex, Ontario, Canada; orifice voltage = 85 volts). Light and heavy chains of the antibody were isolated, purified and refolded according to the procedures described in the preceding references (6-8). Prior to carrying out the degradation reaction of CDRL-1 by the light chain of i41SL1-2 mAb (i41SL1-2-L), most glassware, plastic ware and buffer solution were sterilized by heating (180 °C, 2 hr), autoclaving (121 °C, 20 min) or passing through a 0.2 µm sterilized filter. Manipulations in the experiment were mostly performed in a safety cabinet to avoid contamination from the air. The degradation reaction was carried out in 7.5 mM phosphate containing 5% DMSO, 9 mM HEPES (pH 7.1) at 25 °C. For monitoring of the reaction, 20 µl of the reacting solution was injected into the reversed phase HPLC (Jasco,

Tokyo, Japan) under the isocratic condition with the column temperature of 40 °C.

Messenger RNA was isolated from the hybridoma secreting i41SL1-2 mAb using a mRNA purification kit (Amersham Pharmacia Biotech UK Ltd, UK). Then the cDNAs of the heavy and light chain were synthesized by first-strand cDNA Synthesis Kit (Life Science Inc., FL, USA). Sequencing was carried out using the Auto Read Sequencing Kit (Amersham Pharmacia Biotech) and an automated DNA sequencer (ALF II, Amersham Pharmacia Biotech). The computational analysis was performed by a work station (Silicon Graphics Inc., PA, USA) and the software AbM (Oxford Molecular Ltd., Oxford, UK) which is for building up the three dimensional molecule. The resultant PDB data was applied to minimize the total energy by using the software Quanta 96 (Molecular Simulations Inc., CA, USA). This software uses the algorithm CHARMM for the energy minimization of a molecule (13). For the graphics of the structure, the software Protein Adviser Ver. 3.5 (FQS Ltd., Fukuoka, Japan) was employed.

Results and Discussion

The established i41SL1-2 mAb could specifically bind to the antigenic peptide (CDRL-1-Cys) but not cross-react with other peptides (gp41 peptide, gp120 V3 loop peptide of HIV-1 and VIP) and proteins (HSA, OVA, BSA, p24 of HIV-1). The apparent affinity constants of the intact i41SL1-2 mAb, its heavy and light chain for the antigenic peptide were evaluated by using ELISA. The values were 3.6×10^9, 2.7×10^7, 1.8×10^6/M, respectively. The affinity of the heavy chain was lower than the intact antibody by about one hundred fold. That of the light chain was also lower than the heavy chain by about ten fold. These are normal values as observed when the subunits were isolated from the intact antibody (8).

Sequencing of the cDNAs of variable region of the heavy and light chain of i41SL1-2 mAb was performed and the amino acid sequences were deduced as presented in Figure 1. Sequence of the light chain showed the highest score to germ line bd2. Molecular modeling was carried out utilizing the deduced amino acid sequences. Figure 2 shows the three-dimensional structure of the variable region of i41SL1-2 mAb. A potential catalytic triad composed of Asp, Ser, and His, which are present as a catalytic site of many serine proteases, is generated in the structure. Interestingly, the three amino acid residues (Asp[1] in FR-1, Ser[27A] in CDR1, His[93] in CDR3) are located in the identical positions as reported by Paul for VIP cleaving catalytic antibody light chain (VIPase) (14). Figure 3 shows the comparison of the amino acid sequences of the light chains between i41SL1-2 and VIP cleaving antibody. This coincidence with respect to the positions of three residues implies that the light chain of i41SL1-2 mAb has an ability to catalytically cleave the antigenic peptide, CDRL-1.

a) heavy chain

```
                        10
Q   V   Q   L   Q   H   S   G   A   E   L   V   R   P   G   S   S   V
CAG GTT CAA CTG CAG CAC TCT GGG GCT GAG CTG GTG AGG CCT GGG TCT TCA GTG
    20                                      31              35
K   V   S   C   K   A   L   G   Y   T   S   T   D   Y   E   I   H   W
AAG GTG TCC TGC AAG GCT TTG GGC TAC ACA TCT ACT GAC TAT GAA ATA CAC TGG
                                                    CDR1
        40                              50      52  A   53
V   K   Q   T   P   V   R   G   L   E   W   I   G   A   I   H   P   G
GTG AAG CAG ACA CCT GTG CGT GGC CTG GAA TGG ATT GGA GCT ATT CAT CCA GGA
                    60              65              70
S   D   V   I   V   Y   N   Q   K   F   K   G   T   A   T   L   T   A
AGT GAT GTT ATT GTC TAC AAT CAG AAG TTC AAG GGC ACG GCC ACA CTG ACT GCA
    CDR2
                        80      82  A   B   C   83
D   K   S   S   S   T   A   Y   M   E   L   R   S   L   T   S   E   D
GAC AAA TCC TCC AGC ACA GCC TAC ATG GAG CTC AGA AGT CTG ACA TCT GAG GAC
            90                  95                  102
S   A   V   Y   Y   C   T   R   E   G   G   S   V   D   Y   V   W   G
TCT GCT GTC TAT TAC TGT ACA AGA GAG GGG GGA TCT GTT GAC TAC GTT TGG GGC
                        CDR3
            110
Q   G   T   L   V   T   V   S
CAA GGG ACT CTG GTC ACT GTC TCT
```

b) light chain

```
D   V   V   M   T   Q   (from amino acid sequence analysis)
D   I   V   M   T   Q   T   Q   L   T   L   T   I   N   I   G   Q   P
GAC ATT GTG ATG ACC CAG ACT CAA CTC ACT TTG ACG ATT AAC ATT GGA CAA CCA
    20              24                  27A  B   C   D   E   28
A   S   I   S   C   K   S   S   Q   S   L   L   D   S   D   G   K   T
GCC TCC ATC TCT TGC AAG TCA AGT CAG AGC CTC TTA GAT AGT GAT GGA AAG ACA
                                    CDR1
    34
Y   L   N   W   L   F   Q   R   P   G   Q   S   P   K   R   L   I   Y
TAT TTG AAT TGG TTG TTC CAG AGG CCA GGC CAG TCT CCA AAG CGC CTA ATC TAT
50                  56              60
L   V   S   K   L   D   S   G   V   P   D   R   F   T   G   S   G   S
CTG GTG TCT AAA CTG GAC TCT GGA GTC CCT GAC AGG TTC ACT GGC AGT GGA TCA
    CDR2
    70                          80
G   T   D   F   T   L   K   I   S   R   V   E   A   E   D   L   G   V
GGG ACA GAT TTC ACA CTG AAA ATC AGC AGA GTG GAG GCT GAG GAT TTG GGA GTT
        89                      97          100
Y   Y   C   W   Q   G   T   H   F   P   L   T   F   G   A   G   T   K
TAT TAT TGC TGG CAA GGT ACA CAT TTT CCT CTC ACG TTC GGT GCT GGG ACC AAG
                    CDR3
            107
L   E   L   R
CTG GAG CTG AGA
```

Figure 1. The nucleotide sequences of heavy and light chain variable region of i41SL1-2 and the deduced amino acid sequences. Underline (=) is the region of the primer. In the light chain, six amino acid residues at N-terminus were determined by amino acid sequence analysis as indicated with bold letters.

204

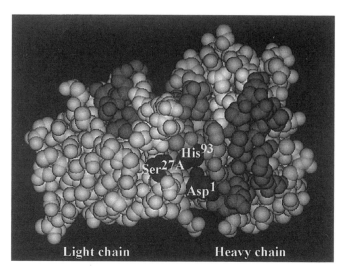

Figure 2. *Structure of the variable region of i41SL1-2 by molecular modeling. A catalytic triad composed of Asp^1, Ser^{27A}, His^{93} is presented in the structure of light chain of i41SL1-2 variable region. The positions of the three amino acid residues are identical to those of the VIP cleaving light chain reported by Paul et al. (14).*

In accordance with the above assumption, the degradation reaction of the CDRL-1 peptide was carried out by using i41SL1-2-L. As shown in Figure 4, the degradation of the CDRL-1 peptide initiated after mixing of i41SL1-2-L and CDRL-1. The CDRL-1 peptide completely disappeared at about 47 hr. In contrast, the CDRL-1 peptide failed to be degraded without i41SL1-2-L. No induction period was observed in this case during the degradation reaction, which is different from the case of super catalytic antibody 41S-2-L (6-8). The HPLC chromatogram is presented in Figure 5. A new peak appeared at the retention time of 19.4 min gradually increased, then reached the maximum and decreased as the reaction time elapsed. The peak was collected and submitted to the analysis of mass spectroscopy, showing m/z ([M+H]$^+$) = 1709.8 and m/z ([M+2H]$^{2+}$) = 855.7. These signals indicate that the molecular weight is 1709.1($\pm$0.4). This is coincident with that of SSKSLLYSNGNTYLY, which is the fragmented peptide of CDRL-1 (RSSKSLLYSNGNTYLY). Consequently, it is suggested that the peptidic bond between Arg^1-Ser^2 of the CDRL-1 was cleaved by i41SL1-2-L. The selective scission for R-X bond is a characteristic feature of serine protease. This fact agrees with the feature of i41SL1-2-L possessing a catalytic triad composed of Asp, His, and Ser.

Though i41SL1-2-L did not cross-react with VIP, the germ line is considered to be the same (bd2) as the VIP cleaving antibody light chain. From the viewpoint of natural catalytic antibody, it is very important to clarify how the germ line which produces the antibody generating a catalytic triad, is recruited.

i41SL1-2 light chain

VIP cleaving light chain

i41SL1-2 light chain

VIP cleaving light chain

```
                              A B C D E
          1         10        20      27 28  30              40                50
i41SL1-2  DVVMTQTQLTLTNIGQPASISCKSSQSLLDSDGKTYLNWLFQRPGQSPKRLIYLVSKLD
VIP       ---------P------SVT----------------HT------T--I----L-----

          60        70        80        90   93        100
i41SL1-2  SGVPDRFTGSGSGTDFTLKISRVEAEDLGVYYCWQGTHFPLTFGAGTKLELR
VIP       -------------------------------------P------Q------G------IK
```

Figure 3. Comparison of the amino acid sequences between i41SL1-2 and the light chain of VIP cleaving antibody (-: indicates the consensus sequence).

206

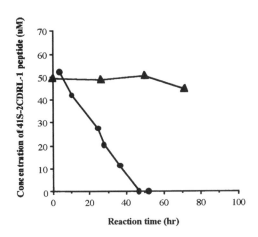

Figure 4. Time course of the degradation of CDRL-1 peptide by the antibody light chain, i41SL1-2-L. CDRL-1, 50 μM; i41SL1-2-L, 0.4 μM; reaction temperature, 25 °C; reaction buffer, 7.5 mM phosphate containing 5% DMSO, 9 mM HEPES (pH 7.1). The CDRL-1 antigenic peptide was immediately degraded by mixing with i41SL1-2-L (●), and completely disappeared at 47 hr. The CDRL-1 antigenic peptide failed to be degraded without i41SL1-2-L (▲). No induction period was observed in this degradation reaction unlike the case of super catalytic antibody 41S-2-L.

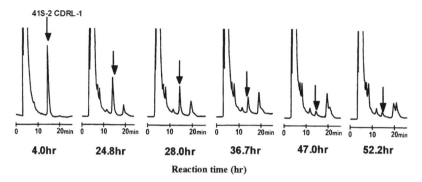

41S-2 CDRL-1

| 4.0hr | 24.8hr | 28.0hr | 36.7hr | 47.0hr | 52.2hr |

Reaction time (hr)

Figure 5. HPLC chromatogram for the degradation of CDRL-1 peptide. Column, Puresil C$_{18}$, Waters; HPLC, Jasco corporation Eluent, 21% acetonitrile including 0.065% TFA under 40 °C at 0.5 ml/min. CDRL-1 peptide at the retention time of 14 min decreased after the mixing with i41SL1-2-L (arrows) and finally disappeared at 47 hr. On the other hand, a new peak at 19.4 min was observed at the reaction time of 24.8 hr, which increased, then reached a maximum and decreased. The peak was identified as being the fragmented peptide SSKSLLYSNGNTYLY by mass spectroscopy.

VIP : NLISNLYKKVAMQKRLRTYNDTFVADSH

CDRL-1 peptide : RSSKS L LY SNGNTYLY

case 1

VIP : NLISNLYKKVAMQKRLRTYNDTFVADSH

CDRL-1 peptide : RSSKS LLYSNGNTYLY

case 2

Figure6. Comparison of the amino acid sequences between CDRL-1 peptide and VIP. Underline (=) is a part of the consensus sequence. Case1 and 2 are the comparisons of the sequences at the different positions.

Figure 6 compares the sequences of VIP with CDRL-1. Three amino acid residues are coincident in case 1 or 2. However, a clear conclusion which kind of feature of immunized antigen recruits bd2 germline cannot be drawn at present time.

Conclusively, this study strongly suggests the way how to explore the natural catalytic antibody by searching the location of Asp, Ser, His residues in light chain after establishing mAbs raised against peptides.

208

Acknowledgements

This study has been supported in part by the Grant-in-Aid (111793007, 12035222, 09217249 and 10180225) for Scientific Research of the Ministry of Education, the Yazaki Memorial Foundation for Science & Technology, the Extensive Research Program of Hiroshima Prefectural Government, and The Association for the Progress of New Chemistry.

References

1. Paul, S.; Volle, D. J.; Beach, C. M.; Johnson, D. R.; Powell, M. J.; Massey, R. J. *Science* 1989, *244*, 1158-1162.
2. Shuter, A. M.; Gololobov, G.V.; Kvashuk, O. K.; Bogomolova, A. E.; Smirnov, I. V.; Gabibov, A. G. *Science* 1992, *256*, 665-667.
3. Sun, M.; Gao, Q. S.; Li, L.; Paul, S. *J. Immunology* 1994, *153*, 5121-5126.
4. Takagi, M.; Kohda, K.; Hamuro, T.; Harada, A.; Yamaguchi, H.; Kamachi, M.; Imanaka, T. *FEBS Letter* 1995, *375*, 273-276.
5. Matsuura, K.; Ikoma, S.; Yoshida, K.; Shinohara, H. *Biochem. Biophy. Res. Commun.* 1998, *243*, 719-721.
6. Uda, T.; Hifumi, E.; Okamoto, Y.; Zhou, Y.; Ishimaru, M. *Clinical Science, Supplement 2* 1998, 429-433. The 12th World AIDS Conference, Geneva (Switzerland).
7. Hifumi, E.; Okamoto, Y.; Uda, T. *J. Biosci. Bioeng.* 1999, *88*, 323-327.
8. Hifumi, E.; Okamoto, Y.; Uda, T. *Appl. Biochem. Biotech.* 2000, *83*, 209-220.
9. Uda, T.; Hifumi, E.; Ohara, K. *Chem. Immunol.* 2000, *77*, 18-32.
10. Vella, C.; Ferguson, M.; Dunn, G.; Meloen, R.; Langedijk, H.; Minor, P. D.; Gen. J. *Virol.* 1993, *74*, 2603-2607.
11. Kennedy, R. C.; Henkel, R. D.; Pauletti, D.; Allan, J. S.; Lee, T. H.; Essex, M.; Dreesman, G. R. *Science* 1986, *231*, 1556-1559.
12. Hifumi, E.; Sakata, H.; Nango, M.; Uda, T. *J. Mol. Catal. A: Chemical* 2000, *155*, 209-218.
13. Anchin, J. M.; Mandal, C.; Culberson, C.; Subramaniam, S.; Linthicum, D. S. *J. Mol. Graphics* 1994, *12*, 257-266.
14. Gao, Q. S.; Sun, M.; Tyutyulkova, S.; Webster, D.; Rees, A.; Tramontano, A.; Massey, R. J.; Paul, S. *J. Biol. Chem.* 1994, *269*, 32389-32393.

Combinatorial Bioengineering and Analytical Methods

Chapter 17

Synthesis and Analysis of Peptide Ligand for Biosensor Application Using Combinatorial Chemistry

Screening and Characterization of 2,3,7-Trichlorodibenzo-*p*-dioxin-Binding Peptide

Yasutaka Morita, Yuji Murakami, Kenji Yokoyama, and Eiichi Tamiya

School of Materials Science, Japan Advanced Institute of Science and Technology (JAIST), 1–1 Asahidai, Tatsunokuchi, Ishikawa 923–1292, Japan

Dioxins and endocrine disrupting chemicals are toxic and bad for health, and they become a one of social problem. The combinatorial synthesis of chemical libraries by the split-and-mix strategy makes it easy to synthesize a large number of peptide libraries and possible to select an optimal peptide that binds with chemicals. In this study, the screening and the characterization of peptide ligands that bind with high affinity to 2,3,7-trichlorodibenzo-*p*-dioxin (2,3,7-TCDD) were carried out. Combinatorial peptide libraries were synthesized by solid phase synthesis and they consisted of hepta-peptide sequences which bind to resin with the C-terminal of the peptide. Five peptides which bound with 2,3,7-TCDD were screened from peptide library, and amino acid sequences of these peptides were investigated. One of the peptides, designed as peptide A, as the concentration of 2,3,7-TCDD increased, the fluorescence decreased almost proportionally in the competitive binding assay. This indicated that the analysis system could be used for detection of 2,3,7-TCDD.

© 2002 American Chemical Society

Introduction

Protein domains or small molecules that mimic protein-protein contact sites are playing an important role in drug discovery, diagnostics, and biotechnology. It is likely that the numbers of small molecules discovered that specifically block protein-protein contact sites will increase as a result of screening synthetic molecular libraries (1). Combinatorial library methods have been widely accepted as useful tools in studies of antigenic determinants, receptor-binding ligands, enzyme substrates and inhibitors. The combinatorial synthesis of chemical libraries by the split-and-mix strategy (2, 3) has been adopted as an efficient means of generating, in relatively few synthetic steps, large numbers of compounds (ligands) for biological evaluation (4). This method makes it easy for us to synthesize a large number of peptide libraries and possible to select an optimal peptide that binds with chemicals.

Dioxins and endocrine disrupting chemicals are toxic and bad for health, and they become one of social problem. There are 210 kinds of dioxin, and they are classified into two large categories: polychlorinated dibenzo-p-dioxins (PCDD) and furans (PCDF), to present time, many endocrine disrupting chemical are found, their total number is more than 140 kinds. The prevention of the discharge and the elimination of these toxic chemicals are desired as soon as possible (5-8).

We studied on the characterization, the binding mechanism with the target, and the application of peptide ligands that bind with high affinity to small molecular chemicals to a biosensor. In this study, we report the properties of peptides that sense the one of the dioxin, 2,3,7-trichlorodibenzo-p-dioxin (2,3,7-TCDD). 2,3,7-TCDD is not so much toxic in comparison to 2,3,7,8-tetrachlorodibenzo-p-dioxin (2,3,7,8-TCDD) (9, 10). It is reported that the antibody for detection of 2,3,7-TCDD has approximately 30% cross-reactivity for 2,3,7-TCDD (11). So, 2,3,7-TCDD was used for target to detect the dioxin, because it is not so much dangerous and has relatively similar structure from the point of view of a research of the anti dioxin antibody.

Materials and methods

Synthesis of peptide library

Combinatorial peptide libraries used in this study were synthesized by solid phase synthesis and they consisted of hepta-peptide sequences which bind to the resin with the C-terminal of the peptide. Aminomethylated polystyrene HL beads (Nova) was used for the resin of peptide synthesis. The diameter of the resin is approximately 100 μm. The combinatorial synthesis of peptide libraries by the split-and-mix strategy (2, 12) was adopted as an efficient means of generating using peptide synthesizer (model: PSSM-8, Shimadzu). Amino acid alignments of CDR in antibodies for small compounds were referred to select amino acid to synthesize combinatorial peptide libraries. Fmoc-Arg (Pmc) -OH, Fmoc-Asn (Trt) -OH, Fmoc-Asp (OtBu) -OH, Fmoc-Gly-OH, Fmoc-Ser (t-Bu) -OH, Fmoc-Thr (t-Bu) -OH and Fmoc-Tyr (t-Bu) -OH were used to synthesize the combinatorial peptide libraries. After synthesis of the peptide library, the protected amino acid radicals were removed from peptide by trifluoroacetic acid.

Screening of 2,3,7-TCDD binding peptide

The combinatorial peptide libraries was mixed with 0.1% (w/v) horseradish peroxidase (HRP) conjugated 2,3,7-TCDD (Stratigic Diagnostic Inc.) in PBS buffer (137 mM NaCl, 3 mM KCl, 8 mM Na_2HPO_4, 2 mM KH_2PO_4 in distilled water, pH 7.3) at room temperature (approximately 25 degree C) for 30 min. Beads were washed with PBS buffer and then enzyme assay was done by adding 1% H_2O_2 and Amplex^{TM} red (10-acetyl-3,7 dihydroxyphenoxazine) (Molecular Probe) in the micro-chamber array (13). Beads that bind with HRP-2,3,7-TCDD were detected by the fluorescence of the assay product (resorufine) (Ex: 573 nm, Em: 590 nm) (Fig. 1) via a fluorescence microscope, and then picked up using a micro-manipulator. These beads were applied to a protein sequencer for the determination of amino acid alignment one by one. The amino acid sequences were determined by the Edman degradation method on an automated pulsed liquid-phase protein-peptide sequencer (model: 476A, PE biosystems). Amino acid alignments of the dioxin binding peptide and their binding properties were determined.

Characterization of screened peptide

The following chemical; biphenyl, dichlorobenzene, diphenylene dioxide, diphenylene oxide, naphthalene, and trichloroethylene were used for the binding inhibitors between 2,3,7-TCDD binding peptide and 2,3,7-TCDD (Fig. 2). The competitive binding assay was carried out as follow (14). The chemical inhibitor and HRP conjugated 2,3,7-TCDD were mixed with the screened peptide on beads. The final concentration of inhibitor and HRP conjugated 2,3,7-TCDD were 16.7 µM and 0.1% respectively. 16.7 µM 2,3,7-TCDD was used for positive control instead of inhibitors. After the incubation, beads were wash with PBS buffer and the enzyme assay was performed by the former method. The fluorescence was observed via a fluorescence microscope, and analyzed using IPLab spectrum system (Scanalytics, Inc.). The effect of the concentration of 2,3,7-TCDD was investigated by competitive binding assay. 0.5 to 500 ppb (1.7-1700 nM) concentration of 2,3,7-TCDD solution were prepared and used for this experiment. The properties of 2,3,7-TCDD binding peptides were compared with the peptide CDR2 (Gly-Ser-Ile-Asn-Pro-Arg-Asn-Gly-Gly-Thr-Tyr-Tyr-Asn-Glu-Arg-Phe-Lys) derived from the anti dioxin antibody (15).

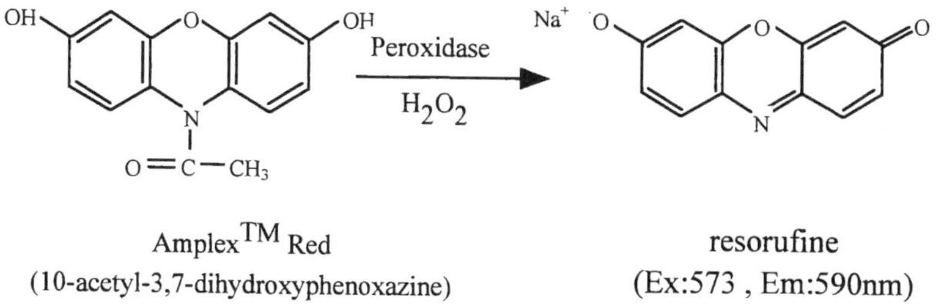

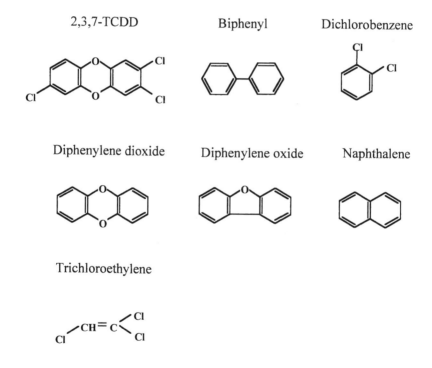

Amplex™ Red
(10-acetyl-3,7-dihydroxyphenoxazine)

resorufine
(Ex:573 , Em:590nm)

Figure 1. Reaction of Amplex™ Red by peroxidase with H₂O₂.

2,3,7-TCDD

Biphenyl

Dichlorobenzene

Diphenylene dioxide

Diphenylene oxide

Naphthalene

Trichloroethylene

Figure 2. Structures of 2,3,7-TCDD and the binding inhibitors between 2,3,7-TCDD binding peptide and 2,3,7-TCDD used in this study.

214

Results and Discussion

Sequence of 2,3,7-TCDD binding peptide

The combinatorial peptide libraries synthesized by solid phase synthesis and used in this work consisted of seven-residue peptide sequences which bound to resin (approximately 100 μm polystyrene beads) with the C-terminal of the peptide. The combinatorial synthesis of peptide libraries by the split-and-mix strategy was adopted as an efficient means of generating. Beads that bound with HRP conjugated 2,3,7-TCDD were observed via a fluorescence microscope and then picked up using a micro-manipulator one by one. These beads were applied to a protein sequencer for the determination of amino acid alignment. Amino acid alignments of the 2,3,7-TCDD binding peptide were investigated as follow; peptide A: Gly-Arg-Asn-Tyr-Gly-Arg-Gly, peptide B: Gly-Asn-Thr-Gly-Tyr-Arg-Gly, peptide C: Gly-Gly-Asp-Arg-Gly-Tyr-Gly, peptide D: Asn-Thr-Asp-Asn-Arg-Asn-Thr, peptide E: Tyr-Tyr-Asp-Gly-Asp-Gly-Gly.

Effect of inhibition between the peptide and 2,3,7-TCDD

The screened peptides really bind with HRP conjugated 2,3,7-TCDD were confirmed using synthesized peptide in the same way (data not shown). Furthermore, the screened peptides and peptide CDR2 were not bound with just HRP (data not shown). In the inhibition test, diphenylene oxide and diphenylene dioxide inhibited the binding between peptide A and 2,3,7-TCDD (Fig. 3A). The reason that peptide A bound with diphenylene oxide and diphenylene, may cause to the similarity of these chemicals with 2,3,7-TCDD. Neither peptide B (Fig. 4A) nor peptide CDR2 (Fig. 5A) were inhibited by the inhibitors.

Detection of 2,3,7-TCDD by competitive binding assay

In the competitive binding assay, fluorescence intensity increased with the decreasing concentration of 2,3,7-TCDD. It revealed that peptide CDR2 detected 2,3,7-TCDD. However, standard error increased with the decreasing concentration of 2,3,7-TCDD (Fig. 5B). The same situation of the detection of 2,3,7-TCDD was observed in the peptide B (Fig. 4B). In the case of peptide A, the fluorescence intensity was increased with the decreasing concentration of 2,3,7-TCDD, and the fluorescence intensity was roughly proportional to the concentration of 2,3,7-TCDD from 0.5 to 500 ppb (Fig. 3B). It means that peptide A can be used to detect 2,3,7-TCDD for the dioxin sensor. Competitive binding assay revealed that screened peptide A, peptide B, and peptide CDR2 can sense 2,3,7-TCDD. The most superior of the 2,3,7-TCDD binding peptide

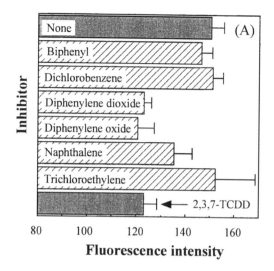

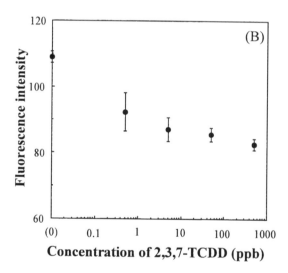

Figure 3. Effect of inhibition between peptide A and 2,3,7-TCDD bound with HRP by chemical compounds (A) and detection of 2,3,7-TCDD by competitive binding assay (B).

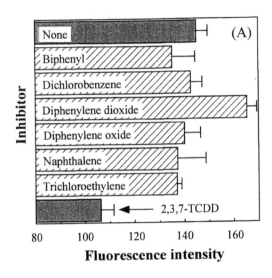

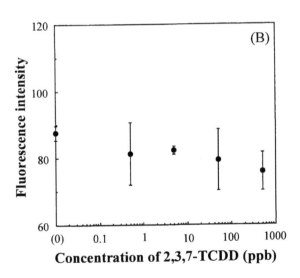

Figure 4. Effect of inhibition between peptide B and 2,3,7-TCDD bound with HRP by chemical compounds (A) and detection of 2,3,7-TCDD by competitive binding assay (B).

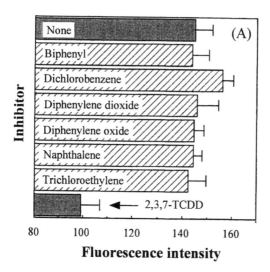

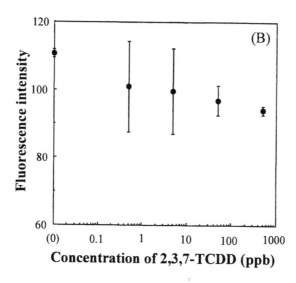

Figure 5. Effect of inhibition between peptide CDR2 and 2,3,7-TCDD bound with HRP by chemical compounds (A) and detection of 2,3,7-TCDD by competitive binding assay (B).

218

among three kinds of peptides was peptide A at the point of sensitivity. The fluorescence intensities were slightly different between graph (A) and (B) in Fig. 3 - 5, although both methods were almost same. Furthermore, errors increased with decreasing concentration of 2,3,7-TCDD in graph (B) of Fig. 3-5. These results may be caused by low affinities of these peptides.

We indicate the capacity of these peptides for biosensor to sense small molecules like 2,3,7-TCDD. In future, we try to modify these peptides to have high affinity for 2,3,7-TCDD.

References

1. Tian, S.-S.; Lamb, P.; King, A.G.; Miller, S.G.; Kessler, L.; Luengo, J.I.; Averill, L.; Johnson, R.K.; Gleason, J.G.; Pelus, L.M.; Dillon, S.B. Rosen, J. A small, nonpeptidyl mimic of granulocyte-colony-stimulating factor. *Science* **1998**, 281, 257-259.
2. Lam, K.S.; Salmon, S.E.; Hersh, E.M.; Hruby, V.J.; Kazmierski, W.M.; Knapp, R.J. A new type of synthetic peptide library for identifying ligand-binding activity. *Nature* **1991**, 354, 82-84.
3. Houghten, R.A.; Pinilla, C.; Blondelle, S.E.; Appel, J.R.; Dooley, C.T.; Cuervo, J.H. Generation and use of synthetic peptide combinatorial libraries for basic research and drug discovery. *Nature* **1991**, 354, 84-86.
4. Lam K.S.; Lebl, M.; Krchnak, V.; *Chem. Rev.* **1997**, 97, 411.
5. Clement, R.E. Ultratrace dioxin and dibenzofuran analysis: 30 years of advaces. *Anal. Chem.* **1991**, 63, 1130-1139.
6. Stone, R. New seveso findings point to cancer. *Science* **1993**, 261, 1383.
7. Stone, R. Toxicologists-and snow-descend on new orleans. *Science* **1993**, 260, 30-31.
8. Gibbons, A. Dioxin tied to endometriosis. *Science* **1993**, 262, 1373.
9. Ahlborg, U.G., Nordic risk assessment of PCDDs and PCDFs. *Chemosphere* **1989**, 19, 603-608.
10 Kutz, F.W.; Barnes, D.G; Bottimore, D.P.; Greim, H.; Bretthauer, E.W., The international toxicity equivalency factor (I-TEF) method of risk assessment for complex mixture of dioxins and related compounds, *Chemosphere* **1990**, 20, 751-757.
11. Sugawara, Y.; Gee, S.J.; Sanborn, J.R.; Gilman, S.D.; Hammock, B.D., Development of a highly sensitive enzyme-linked immunosorbent assay based on polyclonal antibodies for the detection of polychlorinated dibenzo-*p*-dioxin. *Anal. Chem.* **1998**, 70, 1092-1099.
12. Furka, A.; Sebestyen, F.; Asgedom, M.; Dibo, G. General method for rapid synthesis of multicomponent peptide mixture. *Int. J. Pept. Protein Res.* **1991**, 37, 487-493.

13. Nagai, H.; Murakami, Y.; Morita, Y.; Yokoyama, K.; Tamiya, E. Development of a microchamber array for picoliter PCR. *Anal. Chem.* **2000**, 73, 1043-1047

14. Murakami, Y.; Idegami, K.; Nagai, H.; Kikuchi, T.; Morita, Y.; Yamamura, A.; Yokoyama, K.; Tamiya, E. Application of micromachine techniques to biotechnological research, *Material Science and Engineering C* **2000**, 12, 67-70.

15. Recinos, A.3rd; Silvey, K.J.; Ow, D.J.; Jensen, R.H.; Stanker, L.H. Sequences of cDNAs encoding immunoglobulin heavy- and light-chain variable regions from two anti-dioxin monoclonal antibodies. *Gene* **1995**, 158, 311-312.

Chapter 18

Construction of a Combinatorial Protein Library Displayed on Yeast Cell Surface and Its Application to Molecular Breeding

Wen Zou, Mitsuyoshi Ueda*, and Atsuo Tanaka

Department of Synthetic Chemistry and Biological Chemistry, Graduate School of Engineering, Kyoto University, Kyoto 606–8501, Japan

A combinatorial protein library was constructed which could be displayed on the cell surface of *Saccharomyces cerevisiae* by a multi-copy cassette vector. An improved method of DNA random priming was used to generate a large pool of random DNA fragments for increasing the variety. As one of the application, an organic solvent-tolerant yeast strain was obtained by screening of this library with *n*-nonane. Many organic solvents are highly toxic for living organisms because they accumulate in and disrupt cell membranes. Until now, there are few reports on eukaryotic organisms that can adapt to and survive these antimicrobial agents. The acquisition of *n*-nonane-tolerance was proved to be due to the existence of transformed plasmids. The displaying of combinatorial libraries on yeast cell surface might open a novel door for molecular breeding of yeast.

Cell surface is crucial to the life of the cells, including both prokaryotic and eukaryotic cells, because it is a functional interface between the inside and outside of cells. In yeast cells, which are considered as model eukaryotic cells (*1*), surface proteins are closely related to most of the cell surface functions (*2*).

© 2002 American Chemical Society

These structural and functional aspects of the yeast cell surface have practical implications in yeast biotechnology (*3, 4*).

A eukaryotic expression system, using *Saccharomyces cerevisiae* as the host organism, which possesses protein folding and secretary machinery strikingly homologous to that of mammalian cells, is becoming more and more important to express foreign genes. Pioneering research by Dr. Gunter Blobel, the winner of 1999 Nobel Prize in Physiology or Medicine, and his associates revealed that proteins could be directed to many distinct addresses within cells by signal sequences (*5*). This great discovery made it possible to target or locate interested molecules to the cell wall. Therefore, yeast cell surface could be renovated to obtain some new functions by taking advantage of the known transport mechanism of proteins to the cell surface.

Organic solvent tolerance is a cell surface-related function, which is expected to have a wide application in bioprocesses. Until now, in eukaryotic organisms, there has been no genetic information on the mechanism of organic solvent tolerance. In these cases, a random combinatorial protein library, which can be displayed on yeast cell surface, will be helpful, because renovating the cell surface can possibly create some new phenotype strains. We have constructed a combinatorial random protein library that could be displayed on yeast cell surface with the technology of cell surface engineering, and as an application, an *n*-nonane-tolerant yeast strain was obtained by screening this library. The tolerance was confirmed to be dependent on the displaying of the inserted random protein on the yeast cell surface.

Technology of Cell Surface Engineering

Cell surface is related to many cell functions both in prokaryotic and eukaryotic cells, so utilization of cell surfaces is attractive for many applications in microbiology and molecular biology. Based on the theory of signal peptides (*5*), it will be possible to transfer and display various heterologous proteins on the cell surface of microorganisms, which is useful for the segregation of produced polypeptides, construction of microbial biocatalysts, whole-cell adsorbents, and live vaccines.

The surface-expression systems were initially reported from Smith's laboratory, showing that peptides could be fused to the docking proteins (pIII) of a filamentous phage, without affecting its ability to infect *Escherichia coli* (*6*). The development of phage display system (*7*) has already facilitated the isolation of specific ligands, antigens, and antibodies from complex libraries by panning over immobilized substrate (*8*). No sooner had the phage display system been invented than bacterial surface presentation of proteins was achieved. Comparing to phage, bacterial surface may be more suitable for displaying large numbers of proteins. Gram-negative bacterium has the outer membrane at the

222

outmost cell surface, and the heterologous proteins can be displayed by fusing to surface-exposed terminus of the outer membrane proteins (9). Until now, a number of proteins have already been displayed (10,11). Lipoproteins (12), fimbriae (13), and flagella proteins (14) have also been used to immobilize heterologous proteins on the cell surface. Gram-positive bacteria have only recently been taken into consideration for surface display purposes, in which the surface localization of heterologous proteins was achieved by making use of protein A localized on the cell wall (15, 16).

Safe is required when the surface display system is intended to be applied in bio-industrial processes for foods, beverages, medicines and so on. In this case, yeast S. cerevisiae is the most suitable microorganism, which has a GRAS (generally regarded as safe) status, and can be legally used for food, beverage and pharmaceutical production (1). In addition, yeast is an effective host for genetic engineering, since it enables folding and glycosylation of eukaryotic heterologous proteins expressed and is easy to handle with ample genetic techniques.

The yeast cell surface display system was well established in our laboratory (17-19). Until now, many proteins have already been displayed on the yeast surface successfully, including a wide range of molecular mass (20, 21). It has already been confirmed as an efficient system and has a wide application in many biological and biomedical fields.

Surface display expression system for yeast has primarily relied on the fusion of passenger proteins to agglutinin, a protein involved in cell adhesion. α-Agglutinin, one of the glucanase-extractable mannoprotein of S. cerevisiae, was predicted to have 650 amino acids before processing, and possesses a glycosylphosphatidylinositol (GPI) anchor attachment signal at the C-terminal that is required for anchorage on the cell surface (22, 23). In detail, DNA fragments encoding proteins to be displayed are fused to the gene encoding C-terminal 320 amino acids of the α-agglutinin involving the GPI anchor attachment signal sequence. In yeast cells, when C-terminal 320 amino acids of the α-agglutinin is transferred to the outmost surface of the cell wall, the inserted proteins which were fused to α-agglutinin were also displayed on the yeast cell surface (18, 19). The newly displayed proteins will be expected to renovate the yeast cell structures, and will possibly endow the yeast host cells with some new functions and phenotypes.

Combinatorial Protein Library Displayed on Yeast Cell Surface

In the absence of quantitative and computational structure-function relationships for proteins, rational approaches to mutagenesis have limited potential for success in rapidly altering protein molecular properties to meet predefined criteria. An alternative strategy, construction and selection of random-mutated combinatorial libraries, has yielded numerous successes (24). *In vitro* selection from molecular libraries has rapidly come of age as a protein-engineering tool. In a library, the massive number of variants that can be simultaneously surveyed using either chemical or biological approaches have a key consequence for protein engineering. It makes practical to mutate multiple residues of a protein simultaneously, enabling some complex, non-additive combinatorial effects to be obtained. Then, changing proteins properties may rely less on a detailed molecular understanding of its function. Biological libraries should, therefore, offer a powerful way to improve protein function.

As a combination of both library method and yeast surface display technique, we have constructed a combinatorial protein library that could be displayed on the yeast cell surface.

Vector Construction for Displaying on Yeast Cell Surface

A multi-copy plasmid, designated as pMSRH, was constructed to create a combinatorial random protein library that could be displayed on the yeast cell surface based on the technology of cell surface engineering (Fig. 1A). The glyceraldehyde-3-phosphate dehydrogenase (GAPDH) promoter and the gene of the secretion signal sequence of glucoamylase were applied to express and secrete the following fused random protein. As shown in Fig. 1B, the random protein-encoding DNA fragments were fused to the gene encoding the C-terminal 320 amino acids of the α-agglutinin which has a GPI anchor attachment signal sequence required for anchorage on the yeast cell surface. Here, the RGS (His)$_6$ short DNA fragment was used as an epitope to check the subcellular location of the displayed random proteins by fluorescent labeling of the immuno-reaction.

Generation of Random DNA Pool by DNA Random Priming

As shown in Figure 2, we started from the total RNA of *S. cerevisiae* MT8-1 and used the cDNA as the template, following which the method of DNA random priming was applied to generate a pool of random fragments for library construction. These DNA fragments were fused to the gene encoding C-terminal 320 amino acids of α-agglutinin involving a GPI anchor attachment signal sequence. After transformed into yeast cells, the inserted proteins were transferred to the outmost surface of yeast cell wall by the anchorage of α-agglutinin if they were fused in-frame.

The method of DNA random priming was an improved method of random-

A

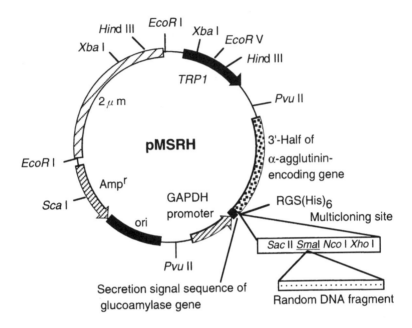

B

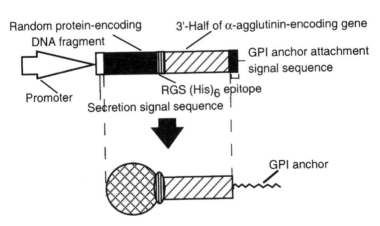

Figure 1. Multicopy cassette vector pMSRH (A) and strategy for surface display on yeast cells (B).

priming recombination (RPR). RPR is a simple and efficient method for *in vitro* mutagenesis and can recombine polynucleotide sequences. Therefore, it is considered as a very flexible and easy way on generating mutant libraries for directed evolution. Different with RPR, DNA random priming produced a lot of fragments with different length and different recombination frequencies. The reaction conditions, such as length and concentration of random primers, could be manipulated to achieve the desired mutagenic rate and recombination frequency.

Cell Surface Display of the Constructed Combinatorial Protein Library

About 4×10^4 yeast single colonies were obtained in the library. The transformed plasmids of 10 random selected colonies were isolated, and the inserted random fragments were sequenced and compared with *S. cerevisiae* Genome Database (data were not shown here). It was found that the recombination frequencies varied between 0%-1% based on the experimental conditions used in this work.

In order to confirm the cell surface display, some of the colonies were randomly selected from the constructed library, and immunofluorescent-labeled (25) with the antibody against RGS(His)$_4$ as the primary antibody at a dilution rate of 1:1000 and fluorescein isothiocyanate (FITC)-conjugated anti-mouse IgG diluted at 1:300 as the second antibody. The labeled cells were observed under a microscope. From these results, it was confirmed that the inserted in-frame random proteins were successfully displayed on the yeast cell surface.

Therefore, by using the method of DNA random priming, we could generate a vast pool of DNA fragments with different length and various structures for the combinatorial library. Proteins displayed on the yeast cell surface led to a great contribution to yeast breeding (18,19). When these fused random proteins were displayed on yeast cell surface, their intrinsic localization, composition and conformation may somewhat be changed, and these changes will increase the variability of the library and enrich the selectable phenotypes.

It seems quite expectable that some new functional proteins can be screened from this combinatorial random protein library. The following shows an example to screen random proteins that can endow organic solvent-tolerance to yeast cells.

Organic Solvent-Tolerant Yeast Strains

Organic solvents are already used widely in the biotransformations with enzymes (26). The use of organic solvents has also several advantages in the application of whole-cell systems. The solvents can increase the solubility of

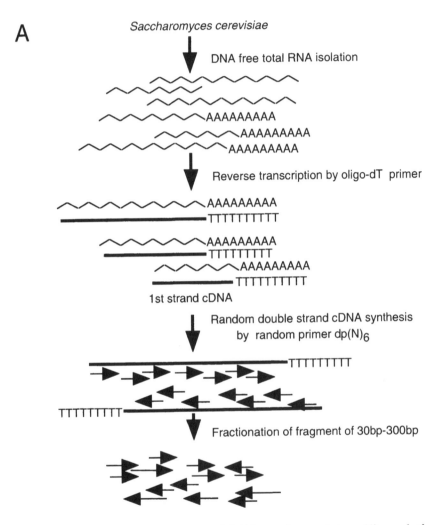

Figure 2. Scheme of the method of DNA random priming (A), and the construction of a random combinatorial protein library to be displayed on yeast cell surface (B) from a pool of random DNA fragments generated by the method.

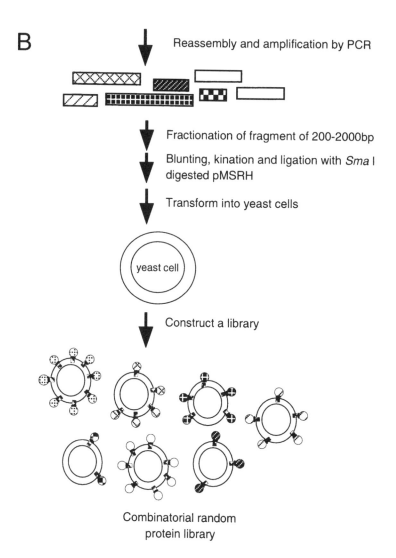

Combinatorial random
protein library

Figure 2. *Continued.*

poorly water-soluble substrates or products. Using a second phase of an organic solvent, products can be extracted continuously from the aqueous reaction system. This process enables not only the reduction of inhibitory effects caused by the product but also a much easier recovery with positive effects on the costs for the downstream processing. However, Organic solvents, like alcohol, aromatics and phenols, are classical antimicrobial agents (27). Most of the hydrophobic organic solvents, such as toluene, are toxic for living organisms because they accumulate in and disrupt cell membrane. The problems resulting from the toxicity of organic solvents to whole cells are still severe drawbacks in biocatalysis (28).

Therefore, microorganisms that can adapt and survive in the presence of organic solvents are of great interest. Recently, some strains of *Pseudomonas aeruginosa, P. fluorescens, P. putida,* and *E. coli* have been found to be tolerant to toxic organic solvents (29-32). But until now, there has been no report on eukaryotic microorganisms with organic solvent tolerance except for the strain KK-21 of *S. cerevisiae,* which is tolerant to isooctane (2,2,4-trimethylpentane) (33). The strain KK-21 is a spontaneous mutant isolated from commercial dry yeast, which was utilized as an immobilized biocatalyst in the double entrapment of calcium alginate and polyurethane for long-term stereoselective reduction of a koto ester in isooctane (34). There is no genetic information about this tolerance.

Cell Surface-Related Mechanism of Tolerance

Research has begun to uncover the mechanisms responsible for the unique property of solvent tolerance since the first solvent-tolerant strain was isolated. Different mechanisms have been proposed in prokaryotic cells (35), including the presence of an efflux system localized in the cell envelope (36-38), which was actively decreasing the amount of solvent in the cell and altering the composition of the outer membrane (39-41). In bacteria, as the membrane is the main target of the toxic action of solvents, the researches of mechanisms are concentrated on the changes of the compositions and structures of membrane. In eukaryotic cells, such as yeast cells, things become complicated because of the different cell structure, especially the different composition of cell envelope. Although there is no report on the mechanism of organic solvent-tolerance of yeast, the complete genome sequence of *S. cerevisiae* revealed the presence of numerous ABC protein-encoding genes (42, 43), whose products locate mostly on plasma membrane and are implicated in drug, metal, and stress resistances (44). Therefore, yeast cell surface is expected to have relationships with organic solvent-tolerance.

This could be confirmed by the cell growth of isooctane-tolerant strain KK-211 (one of the single colonies of strain KK-21) in the presence of both hydrophilic and hydrophobic organic solvents. Surprisingly, the tolerance of the

KK-211 strain to hydrophilic DMSO or ethanol was apparently lower than that of its parent strain, although its tolerance to hydrophobic organic solvents was higher than that of its parent strain. These results suggested that the cell surface property might be different in these two yeast strains. To verify this prediction, cell surface affinity to the hydrophobic organic solvent, isooctane, was investigated by observing the adherence of the cells to isooctane droplets in the aqueous phase as previously performed by Aono *et al.* (*45*). Isooctane was emulsified by vigorous mixing in the suspension of yeast cells. In the case of the wild type cells, most of cells adhered to isooctane droplets, while the KK-211 cells rarely adhered. These results suggested that yeast cell surface was closely related with organic solvent-tolerance. Therefore, in order to identify the mechanism related to organic solvent-tolerance and to find a better application of yeast in industrial processes, we planned to screen for a protein responsible for the tolerance from a combinatorial random protein library.

Screening for a Protein Responding to Organic Solvent-Tolerance from the Library

The transformants were screened by replicating on a series of selective plates, including SD-W plates covered with a half volume of *n*-nonane. The plates were sealed with vinyl plastic tape to prevent vaporization of the organic solvent, and incubated at 30℃ for three days. After successive cultivation, one *n*-nonane-tolerant colony, designated as n13, was finally obtained. The parent *S. cerevisiae* MT8-1 and this tolerant strain n13 were cultivated in *n*-nonane-overlaid liquid YPD medium, and grown at 30^0C. n13 grew very well, while MT8-1 did not (Fig. 3). The fact that n13 did not grow in the medium containing *n*-nonane as a sole carbon source indicated that this strain was not *n*-nonane-trophic, but *n*-nonane-tolerant.

Confirmation of the Tolerance Acquisition

The plasmid-dependence of *n*-nonane-tolerance was checked by the two methods. One was that the cells of MT8-1 gained the tolerance against *n*-nonane when transformed by the plasmid isolated from n13; the other was that n13 lost its tolerance against *n*-nonane after the plasmid was deleted from the cells by cultivating n13 in rich medium YPD for a long time. These results suggested that the *n*-nonane-tolerance of n13 was dependent on the transformed plasmid.

The plasmid-born protein was identified to be displayed on the yeast cell surface by immunofluorescent-labeling of RGS(His)$_4$ epitope, as described previously. Papain treatment was also applied to further investigate whether or not the displayed protein on the cell surface was responsive for *n*-nonane-

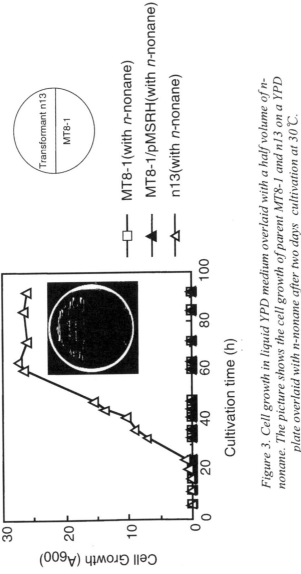

Figure 3. Cell growth in liquid YPD medium overlaid with a half volume of n-nonane. The picture shows the cell growth of parent MT8-1 and n13 on a YPD plate overlaid with n-nonane after two days cultivation at 30°C.

tolerance of n13. Papain was added at different concentrations at the start of cultivation, in the presence of *n*-nonane. At low concentrations, papain did not affect the cell growth of n13. When papain increased to 5 mg in 5ml culture medium, however, the cell growth in the presence of *n*-nonane was significantly inhibited. When papain increased to 10 mg in 5ml culture medium, cell growth was not observed within 50 h of cultivation. On the contrary, in the control experimental group where papain was added at same concentrations in the cultivation, but without the presence of *n*-nonane, even 10 mg papain in 5ml culture medium did not affect severely the cell growth. These experiment data suggested that papain hydrolysis might only destroy the tolerance of n13 cells to *n*-nonane. It, therefore, seemed probable that acquisition of the *n*-nonane-tolerance of n13 was derived from the displayed random protein on the cell surface.

In conclusion, n13 is the first genetically constructed recombinant yeast strain that can tolerate the organic solvent *n*-nonane. It will give us some important information about the organic solvent-tolerance of eukaryotes, which may lead to a much wider application of yeast in industrial bioprocesses in general. At the same time, n13 can be a model for screening of other new phenotypes and novel functional proteins from the combinatorial protein library. We can change the conditions in DNA random priming to meet the expected mutation ratio and screen different molecules by using different probes. Combinatorial random protein library offers a wide application in renovating cell surface, improving protein functions and finding new functional proteins.

References

1. Walker, G. M. *Yeast Physiology and Biotechnology*; John Wiley & Sons: Baffins Lane, Chichester, 1998; 1-50.
2. Fleet, G. H. In *The Yeast;* Rose, A. H. and Harrison, J.S., Ed.; 2nd ed.; Academic Press: London, 1991; Vol. 4: Yeast Organelles, pp 199-277.
3. Schreuder, M. P.; Mooren, A. T. A.; Toschka, H. U.; Verrips, C. T.; Klis, F. M. *Trends Biotechnol.* 1996, *14*, 115-120.
4. Schreuder, M. P.; Dean, C.; Boersma, W. J. A.; Pouwels, P. H.; Klis, F. M. *Vaccine* 1996, *14*, 383-388.
5. Blobel, G. *Cold Spring Harbor Symp. Quant. Biol.* 1995, *60*, 1-10.
6. Scott, J. K.; Smith, G. P. *Science* 1990, *249*, 386-390.
7. Chiswell, D. J.; McCafferty, J. *Trends Biotechnol.* 1992, *10*, 80-84.
8. Hoogenboom, H. R. *Trends Biotechnol.* 1997, *15*, 62-70.
9. Francisco, J. A.; Earhart, C. F.; Georgiou, G. *Proc. Natl. Acad. Sci. USA* 1992, *89*, 2713-2717.
10. Little, M.; Fuchs, P.; Breitling, F.; Dubel, S. *Trends Biotechnol.* 1993, *11*, 3-5.

232

11. Georgiou, G.; Stathopoulos, C.; Daugherty, P. S.; Nayak, A. R.; Iverson, B. L.;Curtiss, R. III. *Nature Biotechnol.* **1997**, *15*, 29-34.
12. Harrison, J. L.; Taylor, I. M.; O Connor, C. D. *Res. Microbiol.* **1990**, *141*, 1009-1012.
13. Hedegaard, L.; Klemm, P. *Gene* **1989**, *85*, 115-124.
14. Newton, S. M. C.; Jacob, C. O.; Stocker, B. A. D. *Science* **1989**, *244*, 70-72.
15. Gunneriusson, E.; Samuelson, P.; Uhlen, M.; Nygren, P.-A.; Stahl, S. *J.Bacteriol.* **1996**, *178*, 1341-1346.
16. Schneewind, O.; Fowler, A.; Faull, K. F. *Science* **1995**, *268*, 103-106.
17. Murai, T.; Ueda, M.; Tanaka, A. *Recent Res. Develop. Microbiol.* **1999**, *3*, 7-22.
18. Ueda, M.; Tanaka, A. *Biotechnol. Adv.* **2000**, *18*, 121-140.
19. Ueda, M.; Tanaka, A. *J. Biosci. Bioeng.* **2000**, *90*, 125-136.
20. Murai, T.; Ueda, M.; Kawaguchi, T.; Arai, M.; Tanaka, A. *Appl.Environ.Microbiol.* **1998**, *64*, 4857-4861.
21. Murai, T.; Ueda, M.; Shibasaki, Y.; Kamasawa, N.; Osumi. M.; Imanaka, T.;Tanaka, A. *Appl. Microbiol. Biotechnol.* **1999**, *51*, 65-70.
22. Lipke, P. N.; Wojciechowicz, D.; Kurjan, J. *Mol. Cell. Biol.* **1989**, *9*, 3155-3165.
23. Wojciechowicz, D.; Lu. C.-F.; Kurjan, J.; Lipke, P. N. *Mol. Cell. Biol.* **1993**, *13*, 2554-2563.
24. Boder, E. T.; Wittrup. K. D. *Nature Biotechnol.* **1997**, *15*, 553-557.
25. Kobori, H.; Sato, M.; Osumi, M. *Protoplasma* **1992**, *167*, 197-204.
26. Carrea, G.; Ottolina, G.; Riva, S. *Trends Biotechnol.* **1995**, *13*, 63-70.
27. Lucchini, J. J.; Corre, J.; Cremieux, A. *Res. Microbiol.* **1990**, *141*, 499-510.
28. Salter, G. J.; Kell, D. B. *Crit. Rev. Biotechnol.* **1995**, *15*, 139-177.
29. Aono, R.; Aibe, K.; Inoue, A.; Horikoshi, K. *Agric. Biol. Chem.* **1991**, *55*, 1935-1938.
30. Aono, R.; Ito, M.; Horikoshi, K. *Biosci. Biotechnol. Biochem.* **1992**, *56*, 145-146.
31. Inoue, A.; Horikoshi, K. *Nature* **1989**, *338*, 264-265.
32. Nakajima, H.; Kobayashi, H.; Aono, R.; Horikoshi, K. *Biosci. Biotechnol. Biochem.* **1992**, *56*, 1872-1873.
33. Kawamoto, T.; Kanda, T.; Tanaka, A. *Appl. Microbiol. Biotechnol.* **2001**, *55*, 476-479.
34. Kanda, T.; Miyata, N.; Fukui, T.; Kawamoto, T.; Tanaka, A. *Appl.Microbiol.Biotechnol.* **1998**, *49*, 377-381.
35. Isken, S.; de Bont, J. A. M. *Extremophiles* **1998**, *2*, 229-238.
36. Isken, S.; de Bont, J. A. M. *J. Bacteriol.* **1996**, *178*, 6056-6058.
37. Ramos, J. L.; Duque, E.; Rodoriguez-Herva, J. J.; Godoy, P.; Haidour, A.; Reyes, F.; Fernandez-Barrero, A. *J. Biol. Chem.* **1997**, *272*, 3887-3890.
38. White, D. G.; Goldman, J. D.; Demple, B.; Levy, S. B. *J. Bacteriol.* **1997**, *179*, 6122-6126.

39. Ingram, L. O. *Appl. Environ. Microbiol.* **1977**, *33*, 1233-1236.
40. Pinkart, H. C.; Wolfram, J. W.; Rogers, R.; White, D. C. Appl. Environ. *Microbiol.* **1996**, *62*, 1129-1132.
41. Weber, F. J.; Isken, S.; de Bont, J. A. M. *Microbiology* **1994**, *140*, 2013-2017.
42. Decottignies, A.; Goffeau, A. *Nature Genet.* **1997**, *15*, 137-145.
43. Kuchler, K.; Egner, R. In *Unusual Secretary Pathways: From Bacteria to Man*; K. Kuchler and B. Rubartelli, Ed.; Landes Bioscience, Austin, 1997; pp.49-85.
44. Bauer, B. E.; Wolfger, H.; Kuchler, K. *Biochim. Biophys. Acta* **1999**, *1461*, 217-236.
45. Aono, R.; Kobayashi, H. *Appl. Environ. Microbiol.* **1997**, *63*, 3637-3642.

Chapter 19

Construction of Surface-Engineered Arming Yeasts with Fluorescent Protein Sensors in Response to Environmental Changes

Seiji Shibasaki[1,2], Mitsuyoshi Ueda[2,*], and Atsuo Tanaka[2]

[1]Department of Applied Chemistry, Kobe City College of Technology,
Kobe 651–2194, Japan
[2]Department of Synthetic Chemistry and Biological Chemistry, Graduate
School of Engineering, Kyoto University, Kyoto 606–8501, Japan

Although biosensors using microorganisms play important roles in bioprocesses, sensing abilities in view of versatility are quite limited because the sensing system depends on only a single metabolic reaction. Moreover, for withdrawing of the information from these systems, there are a lot of steps including invasive methods. Appearance of gene-engineered cells made us possible to draw the several information from cells with non-invasive manner at real time. First of all, *Aequorea victoria* GFP was successfully displayed as its active form on *Saccharomyces cerevisiae* depending on promoters. Secondly, GFP and its variant could be expressed on same cell surface in response to glucose concentration by the different colors. That cell was able to put out the intra- and extracellular informations. Thirdly, quantitative systems of fluorescent protein-displaying cell were examined and confirmed as useful for basic and practical studies. We have developed novel systems of non-invasive sensing of metabolites by using the surface-engineered yeast cells with fluorescent proteins.

© 2002 American Chemical Society

On bioprocesses, high performance is dependent on the abilities of living cells used as the biocatalysts. The data should be taken on-line and real time under non-invasive monitoring concepts. Taking into account these problems, many kinds of sensing systems have been developed to acquire informations on several components in media. For example, a flow injection analysis (FIA) is one of the excellent techniques in the on-line sensing system. In FIA, cell-free medium is continuously withdrawn from the culture to monitor concentrations of one or more analytes. The analyte is measured after appropriate dilution to adjust its concentration within the sensitivity range of the sensor (1). However, FIA is not suitable for sensing of intracellular concentration of analytes because the process of disruption is indispensable in those cases (2, 3). In bioprocesses, a simultaneous sensing of both intra- and extracellularcomponents with non-invasive manner is very important to optimize the processes.

In order to solve this problem, a system of the sensor which can monitor the concentrations of cytosolic components without disruption of cells must be developed. The construction of the cells which can put out the information about intracellular events at real time will be very valuable. Genetic engineering should be introduced into the cells for endowing such sensing ability.

The first report of the surface display experiment using fusions of the docking proteins and peptides on phage (4) became the origin of the phage display system (5). Also, there are a lot of reports about cell surface display of heterologous proteins on bacterial or eukaryotic cells. Escherichia coli has a membrane protein OmpA on its surface, and display of cellulose binding domain (6) or β-lactamase (7) fused with OmpA were succeeded. The glycoprotein of human respiratory syncytial virus was displayed on Staphylococcus xylosus (8) using cell wall regions X and M from staphylococcal protein A. Besides these examples, it is interesting that an HIV-1-gp41 epitope inserted into the antigenic site B of influenza virus hemagglutinin, expressed on the surface of baculovirus-infected insect cells (9). As more practical system, we have developed cell surface engineering using Saccharomyces cerevisiae since1997 (10, 11).

To display heterologous proteins on the yeast cell wall, "the most outer surface protein", α-agglutinin (12) has been used. α-Agglutinin is composed of a secretion signal region, a binding region, a serine- and threoninre-rich region, and a glycosylphosphatidylinositol (GPI) anchor (13) attachment signal sequence. In the transportation process, GPI-anchored α-agglutinin is sorted to the plasma membrane through exocytosis and then released by phosphatidylinositol-specific phospholipase C (PI-PLC) to the cell wall (13). Finally, α-agglutinin is fixed to cell wall by addition of β-1,6-glucan to the GPI anchor remnant of α-agglutinin (14). So far, bacterial hydrolytic enzymes have been displayed on S. cerevisiae genetically as fusion proteins with the C-terminal half of α-agglutinin (10, 15-17). These cells have been called "arming yeasts". Genetic constructions of this display system require a suitable promoter, a secretion signal sequence and a transcriptional terminator.

Various reporter proteins can be displayed on the surface depending on the environmental changes. If display of the reporter on the cell surface is successful, the cell will tell us the information, "what happens inside or outside of the cell", without disruption of cell. Furthermore, this system will have the advantage that metabolic systems will not be interfered by accumulation of hetelorogous proteins in the cytoplasm. To create such interesting engineered cells, suitable reporters to be displayed are indispensable.

As a reporter of above system, *Aequorea victoria* green fluorescent protein (GFP) is thought to be most suitable. Formerly, other biological probes, radio isotope (RI) has been used to analyze cellular and molecular functions, to perform the DNA sequencing, to monitor metabolisms and so on. At present, RI has been replaced to fluorescent reagents or fluorescent proteins for its convenience to handling in experiment and high sensitivity. GFP is the most powerful reporter of fluorescent compounds because it does not need any co-factor other than molecular oxygen. Also in the sensing at bioprocesses, it will be expected to utilize as reporter protein in the various cells. In *S. cerevisiae*, GFP variants have been expressed successfully (*18-20*). These indicate that GFP has the possibility that it would be displayed on yeast cell surface.

Promoters which can respond to environmental conditions will function as information-receiver element for displaying of GFP. Five years ago, although genomic sequence of *S. cerevisiae* was completely revealed, many transcriptional studies in *S. cerevisiae* had already been reported and the practical studies have been performed. The first published example was the use of *ADH1* promoter for efficient intracellular expression of leukocyte α-interferon (*21*). Although this promoter was derived from yeast its own, several foreign promoters have been available to gene expression in yeast cell (*22, 23*). Right choice of a promoter to display a fluorescent protein will bring about a lot of information for us.

Here we summarize unprecedented sensing and monitoring systems by combining the cell surface display system, functional promoters, and fluorescent proteins. It would be expected that the novel systems offer non-disruptive and real-time sensing or monitoring of living cells.

Display of GFP on Yeast Cell Surface

A visible cell surface reporter of *S. cerevisiae* was molecularly designed as follows (*24*). A secretion signal sequence of the glucoamylase precursor protein (*25*) was first connected to the N-terminal of GFPuv. To locate GFPuv on the cell wall, the C-terminal half sequence of α-agglutinin was tagged to the C-terminal of GFPuv. Finally, a strong glucose-regulative promoter *GAPDH* (glyceraldehyde-3-phosphate dehydrogenase) was placed on the upstream of the secretion signal sequence-encoding gene. The molecular structure of the fused gene is illustrated in Fig. 1A, and the plasmid pICS:GFPuv harboring the fused gene is shown in Fig.1B.

The transformant cells were grown in YPD medium until the exponential growth phase and collected for detection of fluorescence. As a control, the parent

237

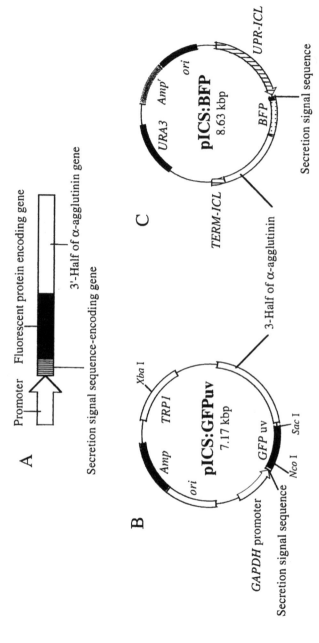

Figure 1. The gene structure for displaying a visible reporter on cell surface of S. cerevisiae (A), the construction of the plasmid pICS:GFPuv (B), and the plasmid pICS:BFP (C).

strain *S. cerevisiae* MT8-1 was grown under the same conditions. GFPuv was functionally expressed in *S. cerevisiae* grown on glucose, but not on ethanol and acetate. Immunofluorescence labeling with rabbit anti-GFP IgG (as the primary antibody) and Rhodamine Red^{TM}-X-conjugated goat anti-rabbit IgG (as the second antibody) which cannot penetrate into the cell membrane was performed to verify the presence and the localization of GFPuv on the cell surface (*24*). Red fluorescence could be observed for the transformant cells, but not for the control cells. Thus, it was confirmed that GFPuv was actually anchored on the cell wall and was able to fold correctly and to form the chromophore even when targeted to the cell wall. Although two extra amino acid residues (Ala-Gly) and eight extra amino acid residues (Gly-Arg-Gly-Asp-Leu-His-Gly-Ser) were introduced between the secretion signal sequence and GFPuv and between C-terminal half of α-agglutinin, respectively, the functional GFPuv was displayed on the cell surface.

To examine whether the fused protein was leaked into the medium or retained by cells, the fluorescence of cells, the medium and each fraction after the subcellular fractionation was monitored using a fluorescence spectrophotometer. Cells were cultivated either in a rich (YPD) medium or in a minimum (SD-W) medium. Cells at the exponential growth phase were collected, followed by isolation of each fraction, including the cell wall fraction after β-1,3-glucanase treatment. The emission of fluorescence was detected only with the β-glucanase extract of the cells harboring pICS:GFPuv (Table I). No fluorescence could be detected from all fractions of control cells. These results revealed that no GFP leaked out from the cells or retained inside cells. Compared to YPD medium, GFP fluorescence was very weak when cells were grown in SD-W (SD not containing tryptophan) medium. The native polyacrylamide gel electrophoresis (PAGE) analysis supported the fluorescent analysis of each cell fraction .

Although the GFP-fused cell wall mannoprotein has been used to detect the localization in the cell wall under the control of the own promoter of the cell wall protein (*26*), it would not be appropriate to utilize as a cell surface reporter because its activation and inactivation are not strongly dependent on the changes in the environment. In our system, a glucose-inducible promoter *GAPDH* directly controlled the expression of the GFP-fused gene. A GPI-anchor attachment signal sequence in the 3'-half of α-agglutinin was used to localize GFP to the cell wall. So far various kinds of promoters with either a positive or a negative response to the intracellular molecular events have been identified and characterized. Most of them could be used to control the display of GFP on the cell surface, which allows us to measure the intracellular concentration of metabolites *in vivo* as the activators or repressors onto the promoters. A variety of GFP-type fluorescent proteins could also be used in these systems to give different colors responding to each molecular event occurring inside and outside cells. It should be pointed out that at present it is not difficult to measure the extracellular concentrations of metabolites or nutrients on-line with current techniques. However, the on-line and *in vivo* detection of intracellular

Table I. Distribution of the expressed fluorescent proteins
in each fraction of *S. cerevisiae*

Cell & Medium	Medium	Cell-free extract	β-Glucanase extract
MT8-1			
YPD medium	0	0	0
SD-W medium	0	0	0
MT8-1/pICS:GFPuv			
YPD medium	0	0	145
SD-W medium	0	0	24.3
MT8-1/pICS:BFP			
YPD medium	0	21.0	128

Note: Units are relative fluorescent unit (RFU).

concentrations of metabolites at the level of a single cell is not still available. Our system may find wide application in bioprocess engineering, e.g., in the form of on-line or *in vivo* monitoring of cell growth and metabolism at the level of a single cell. With the present method, it is possible to monitor the glucose concentration from 0 g/l to 20 g/l.

Sensing of Glucose Concentration

To detect the exhaustion of glucose, the *UPR-ICL* promoter and one of GFP variants, BFP (blue fluorescent protein) were used as the second reporter (*27*). *Candida tropicalis UPR-ICL* promoter can induce the transcription when glucose is depleted, and well function even if in *S. cerevisiae* (*28*). BFP contains three amino acid substitutions, so that the emission is shifted from green to blue (*29*).

For the stable display of BFP, the plasmid pICS:BFP (Fig. 1C) was digested by *Eco*RV and introduced into *S. cerevisiae* MT8-1 strain to be integrated into the chromosome. The obtained transformant was named MT8-1/pICS:BFP strain.

The cells harboring the BFP/α-agglutinin fusion gene were proved to emit blue fluorescence at the stationary growth phase in YPD medium. The strain constructed by integration of pICBS1 as the control (*27*) was hardly labeled. Thus, it was confirmed that the displayed BFP/α-agglutinin fusion protein was anchored on the cell wall of the cells integrating the plasmid pICS:BFP. To determine whether or not BFP was secreted in the culture medium, retained in the cell wall or localized in cytoplasm, cells were cultivated in YPD medium at 30°C for 24 h. Culture medium and cell pellet were separated by centrifugation to measure the fluorescence of BFP in both fractions. Cell pellet was futher treated with laminarinase. The result shown in Table I demonstrated that the strain MT8-1/pICS:BFP had the cell wall-associated BFP without secretion of the protein. A slight fluorescence compared with that of cell wall was detected in the cell free extract.

To shift the expression of GFP to that of BFP, both the GFP/α-agglutinin-enocoding gene under the control of the *GAPDH* promoter and the BFP/α-agglutinin-encoding gene under the control of the *UPR-ICL* promoter, which is functional after exhausting glucose, were integrated into the chromosomes of strain MT8-1. The plasmid pICS:GFPuv (Fig. 1) was digested by *Xba* I and introduced into the MT8-1/ pICS:BFP. A cell of uracil and tryptophan non-auxotrophy was picked up and named MT8-1/GBFP strain. Display of both fluorescent proteins was confirmed by immunofluorescence microscopy under a high and low concentrations of glucose. The fluorescence intensity of GFP or BFP extracted from cell walls were measured at several cultivation times. The relationship between glucose concentration and fluorescence intensity is shown in Fig. 2. Display of GFP on the cell surface was initiated at the early stage of cultivation, while display of BFP started after exhausting glucose. The maximum

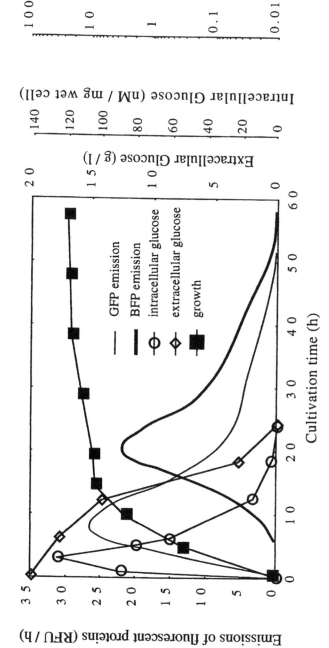

Figure 2. Monitoring of cell growth and glucose concentration by emission from GFP and BFP displayed on the cell surface of S. cerevisiae during cultivation.

RFU of GFP was observed at 7 h cultivation when the glucoes concentration in the medium was 17.2 g/l. For folding and transportation to the cell wall, the delay of the GFP fluorescence might occur. On the other hand, the maximum RFU of BFP was obtained at 19 h cultivation when the glucose concentration in the medium was reduced to 2.5 g/l. This result indicates that GFP was successfully controlled by the *GAPDH* promoter and BFP by the *UPR-ICL* promoter. This system can be applied to the quantitative measurement of produced foreign proteins, and this practical use of the sensing method will be reported elsewhere.

We succeeded in the construction of yeast cells exhibiting the shift of GFP display to BFP display depending on glucose concentration. Display of BFP on the cell surface was confirmed by fluorescence microscopy, confocal laser scanning microscope (CLMS), and the treatment of the cell wall with laminarinase. The cell surface-engineered (arming) cells had an attractive response to glucose concentration. Thus, the combination of the promoters which respond to the change of environmental factors enables the cells to have higher functions on the cell surface with different colors and intensities of fluorescence from the cell surface in response to the change of molecular events inside and outside cells.

Quantitative Analysis of Displayed Protein on Yeast Surface

Visible reporter on the cell surface provides a quantitative evaluation system for cell surface engineering. A variety of heterologous enzymes could be displayed on the surface of the so-called arming yeast cells. *In vivo* quantification of enzyme activity or of the total amounts of proteins displayed on the cell surface is not easy. Although a fluorescent antibody method may be used for this purpose, antibody can only bind to the out-layer proteins displayed on the cell surface. The proteins located in the inner-layer may not have a chance to bind with antibody, and this makes the quantitative assay very difficult. The display of a visible reporter on the cell surface may contribute to solving this problem since it can be directly measured using not only a fluorometer but also a CLSM combined with appropriate image processing (*30*). CLSM has been already used cytochemical evaluation of localization of cell surface protein (*31*).

We selected EGFP to target on the cell surface because it is brighter than native GFP and GFPuv (*32*). The plasmid pIEG1 is a chromosome-integrative type plasmid for expression of the EGFP/α-agglutinin fusion gene containing a secretion signal sequence of glucoamylase under the control of the *GAPDH* promoter. The plasmid pMEG1 was a multicopy plasmid for expression of the same gene.

Fluorescence microscopic observation was carried out to determine whether the transformants emit fluorescence from the cell suface. Cells harboring the plasmid pIEG1 or pMEG1 were cultivated in SD-W liquid medium. As a control, cells harboring the plasmid pICAS1 or pCAS1 were used.

After cultivation, cells were harvested and washed twice with PBS buffer (pH 7.4). The level of fluorescence of cells harboring pMEG1 was higher than that of cells harboring pIEG1. Immunostaining using anti-GFP antibody and Alexa Fluor[TM]546-conjugated goat anti-mouse IgG having different excitation/ emission wavelengths from those of EGFP was performed to determine whether these fluorescences were derived from EGFP.

Emission of fluorescence on the cell surface decreased after 24 h cultivation to the stationary phase. To examine the effect of pH of the medium on the level of emission of EGFP, pH of the SD-W liquid medium was monitored during cultivation. During exponential growth, pH drastically decreased from 5.5 to 3.0, resulting in the decrease of fluorescence. Therefore, we tried to bufferize the medium. The cells cultivated in the SD-W liquid medium adjusted to pH 7 with HEPES (SD-WH7) maintained the intensity of fluorescence for a longer time. The maximal fluorescence was obtained at 24h in SD-WH7 medium. The medium provided us very clear images of CLSM for EGFP displaying cell (*30*) (Fig. 4). There was bud scars as a ring in the photograph vividly. Quantification of fluorescence were also carried out with the cells cultivated in SD-WH7 using two methods.

The number of the EGFP molecules displayed on the surface was calculated by measuring the fluorescence intensity of the cells by a fluorometer (Fig. 3). The following formula obtained from the standard curve was applied for calculation.

The number of the displayed EGFP molecules per cell

$$= RFU_f \times 2.16 \times 10^{11} / 3.33 \times 10^7$$

Where RFU_f (relative fluorescence unit) is an arbitrary unit measured with a fluorometer under the conditions of 1 $OD_{600} = 3.33 \times 10^7$ cells/ml. The results are summarized in Table II.

The number of the EGFP molecules displayed on a cell surface of one cell was also determined by processing the CLSM images (Fig. 4). The following formula was obtained from the calibration curve.

The number of the displayed EGFP molecules per cell = $RFU_c \times 6.16 \times 10^{-9}$
Where RFU_c is the arbitrary unit determined from the analysis of the image of CLSM using the processing software. In processing images from sections taken, signals of intracellular autofluorescences and cytoplasmic fluorescence of EGFP were eliminated and fluorescence intensities only from the cell surface were calculated using the TRI image processing software.

The numbers of the EGFP molecules on the cell surface obtained by two methods were almost similar (Table II). In both cases, the intensity of fluorescence from MT8-1/pMEG1 was 3-4 fold more than that from MT8-1/pIEG1.

We proposed here two methods in the quantification of GFP fused to α-agglutinin displayed on the cell surface of yeast. Ward and Bokman (*33*) suggested that 50-60 % of native GFP denatured by acids (pH 2) would be renatured by shifting pH to 8 in a solution. We carried out such a denaturation-

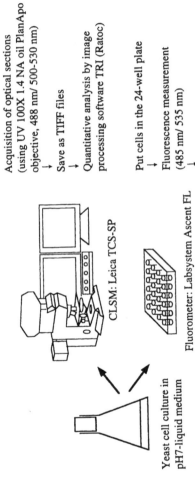

Yeast cell culture in pH7-liquid medium

CLSM: Leica TCS-SP

Fluorometer: Labsystem Ascent FL

Acquisition of optical sections
(using UV 100X 1.4 NA oil PlanApo
objective, 488 nm/ 500-530 nm)
→
Save as TIFF files
→
Quantitative analysis by image
processing software TRI (Ratoc)

Put cells in the 24-well plate
→
Fluorescence measurement
(485 nm/ 535 nm)
→
Subtraction of control cells'
fluorescence intensity from EGFP-
displaying cells' fluorescence intensity

Figure 3. Outline of quantitative analysis using CLSM image processing and fluorometer.

Table II. The number of EGFP molecules displayed on the cell surface

| | Integrative-type transformants | | Multicopy-type transformants | |
	MT8-1/pICAS1	MT8-1/pIEG1	MT8-1/pCAS1	MT8-1/pMEG1
Fluorometric assay	0	$(2.75 \pm 1.66) \times 10^4$	0	$(8.35 \pm 2.89) \times 10^4$
CLSM analysis	0	$(2.45 \pm 1.59) \times 10^4$	0	$(7.50 \pm 2.72) \times 10^4$

Note: Units are molecules per cell.

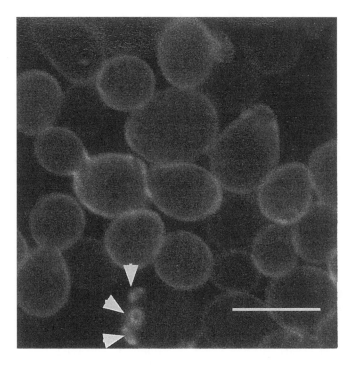

Figure 4. CLSM observation of cell surface of S. cerevisiae. Bar=5 µm. Arrows: bud scars.

246

renaturation test using recombinant EGFP, in which renaturation was possible at pH 7.4 by adding NaOH after denaturation by HCl, and confirmed that EGFP is more sensitive to the change of pH than native GFP as described by Patterson et al. (*34*), although the high intensity of fluorescence of EGFP was obtained. Under the optimum pH, a simple fluorometric assay was examined to quantify the number of the EGFP molecules on the cell surface. CLSM image analysis was also performed using one cell. In both methods, similar results of displayed numbers of the EGFP molecules on the cell surface were obtained as shown in Table II. These results revealed that the fluorometric analysis facilitates the measurement of the amounts of the proteins displayed on the cell surface in taking a consideration of the comparison of easiness for these apparatuses to be operated. Additionaly, display of more pH-sensitive fluorescent protein might produce the cell which can sense pH.

References

1. Bambot, S.B.; Lakowicz, J.R.; Rao, G. *Trends Biotechnol.* **1995**, 13, 106-115.
2. Steube, K.; Spohn, U. *J. Biotechnol.* **1994**, 33, 221-231.
3. Kracke, H.A.; Brandes, L.; Hitzmann, B.; Rinas, U.; Schugerl, K. *J. Biotechnol.* **1991**, 20, 95-104.
4. Scott, J.K.; Smith, G.P. *Science* **1990**, 249, 386-390.
5. Chiswell, D.J.; McCafferty, J. *Trends Biotechnol.* **1992**, 10, 80-84.
6. Francisco, J.A.; Stathopoulos, C.; Warren, R.A.; Kiburn, D.G.; Georgiou, G. *Biotechnology*, **1993**, 11, 491-495.
7. Georgiou, G.; Stephens, D.L.; Stathopoulos, C.; Poetschke, H.L.; Mendenhall, J.; Earhart, C.F. *Protein Eng.* **1996**, 9, 239-247.
8. Nguyen, T.N.; Gourdon, M.H.; Hansson, M.; Robert, A.; Samuelson, P.; Libon, C.; Andreoni, C.; Nygren, P.A.; Binz, H.; Uhlen, M.; Stahl, S. *J. Biotechnol.* **1995**, 42, 207-219.
9. Ernst, W.; Grabherr, R.; Wegner, D.; Borth, N.; Grassauer, A.; Katinger, H. *Nucleic Acids Res.* **1998**, 26, 1718-1723.
10. Murai, T.; Ueda, M.; Yamamura, M.; Atomi, H.; Shibasaki, Y.; Kamasawa, N.; Osumi, M.; Amachi, T.; Tanaka, A. *Appl. Environ. Microbiol.* **1997**, 63, 1362-1366.
11. Ueda, M.; Murai, T.; Shibasaki, Y.; Kamasawa, N.; Osumi, M.; Tanaka, A. *Ann. N.Y. Acad. Sci.* **1998**, 864, 528-537.
12. Terrance, K.; Heller, P.; Wu, Y.-S.; Lipke, P.N. *J. Bacteriol.* **1987**, 169, 475-482.
13. Lu, C.-F.; Kurjan, J.; Lipke, P.N. *Mol. Cell. Biol.* **1994**, 14, 4825-4833.
14. Lu, C.-F.; Montijn, R. C.; Brown, J. L.; Klis, F.; Kurjan, J.; Bussey, H.; Lipke, P. N. *J. Cell Biol.* **1995**, 128, 333-340.

15. Murai, T.; Ueda, M.; Atomi, H.; Shibasaki, Y.; Kamasawa, N.; Osumi, M.; Kawaguchi, T.; Arai, M.; Tanaka, A. *Appl. Microbiol. Biotechnol.* **1997,** 48, 499-503.
16. Murai, T.; Ueda, M.; Kawaguchi, T.; Arai, M.; Tanaka, A. *Appl. Environ. Microbiol.* **1998,** 64, 4857-4861.
17. Murai, T.; Ueda, M.; Shibasaki, Y.; Kamasawa, N.; Osumi, M.; Imanaka, T.; Tanaka, A. *Appl. Microbiol. Biotechnol.* **1999,** 51, 65-70.
18. Bell, P.J.; Davies, I.W.; Attfield, P.V. *Yeast* **1999,** 15, 1747-1759.
19. Frank, P.; Braunshofer, R.C.; Karwan, A.; Grimm, R.; Wintersberger, U. *FEBS Lett.* **1999,** 450, 251-256.
20. Damelin, M.; Silver, P.A. *Mol. Cell* **2000,** 5, 133-140.
21. Hitzeman, R.A.; Hagie, F.F.; Levine, H.L.; Goeddel, D.W.; Ammerer, G.; Hall, B.D. *Nature* **1981,** 293, 717-723.
22. Umemura, K.; Atomi, H.; Kanai, T.; Teranishi, Y.; Ueda, M.; Tanaka, A. *Appl. Microbiol. Biotechnol.* **1995,** 43, 489-492.
23. Walker, G.M. *Yeast Physiology and Biotechnology.* John Wiley & Sons, **1998,** 275.
24. Ye, K.; Shibasaki, S.; Ueda, M.; Kamasawa, N.; Osumi, M.; Shimizu, K.; Tanaka, A. *Appl. Microbiol. Biotechnol.* **2000,** 54, 90-96.
25. Ashikari, T.; Nakamura, N.; Tanaka, N.; Kiuchi, N.; Shibano, Y.; Tanaka, T.; Amachi, T.; Yoshizumi, H. *Agric. Biol. Chem.* **1986,** 50, 957-964.
26. Ram, A.F.J.; Van den Ende, H.; Klis, F.M. *FEMS Microbiol. Lett.* **1998,** 162, 249-255.
27. Shibasaki, S.; Ueda, M.; Ye, K.; Kamasawa, N.; Osumi, M.; Shimizu, K.; Tanaka, A. *Appl. Microbiol. Biotechnol. in press.*
28. Kanai, T.; Atomi, H.; Umemura, K.; Ueno, H.; Teranishi, Y.; Ueda, M.; Tanaka, A. *Appl. Microbiol. Biotechnol.* **1996,** 44:759-765.
29. Mitra, R.D.; Silva, C.M.; Youvan, D.C. *Gene* **1996,** 173, 13-17.
30. Shibasaki, S.; Ueda, M.; Iizuka, T.; Hirayama, M.; Ikeda, Y.; Kamasawa, N.; Osumi M.; Tanaka, A. *Appl. Microbiol. Biotechnol.* **2001,** 55, 471-475.
31. Shibasaki, Y.; Kamasawa, N.; Shibasaki, S.; Zou, W.; Murai, T.; Ueda, M.; Tanaka, A.; Osumi, M. *FEMS Microbiol. Lett.* **2000,** 192, 243-248.
32. Yang, T.T.; Cheng, L.; Kain, S.R. *Nucleic Acids Res.* **1996,** 24, 4592-4593.
33. Ward, W.W.; Bokman, S.H. *Biochemistry* **1982,** 21, 4535-4540.
34. Patterson, G.H.; Knobel, S.M.; Sharif, W.D.; Kain, S.R.; Piston, D.W. *Biophys. J.* **1997,** 73, 2782-2790.

Chapter 20

Sensor Peptides Based on Fluorescence Resonance Energy Transfer

Kenji Yokoyama*, Masayuki Matsumoto, Hideo Ishikawa, Yasutaka Morita, and Eiichi Tamiya

School of Materials Science, Japan Advanced Institute of Science and Technology, 1–1 Asahidai, Tatsunokuchi, Ishikawa 923–1292, Japan
*Correspond author: email: yokoyama@jaist.ac.jp

Sensor peptides based on fluorescence resonance energy transfer (FRET) are described. Prior to developing sensor peptides, FRET-based indicators for the secondary structure of peptides was designed. Helical peptides, $AY(KAAAA)_n KAC$ (n = 1, 2, 3), were selected as the model indicator molecules, and the N-terminal α-amino group and C-terminal cysteine side-chain were modified with rhodamine B sulfonyl chloride and fluorescein maleimide, respectively. The resulting fluorescence spectra showed that FRET from fluorescein to rhodamine B was clearly observed in trifluoroethanol (TFE)-containing buffer, and that fluorescence intensity due to the FRET was reduced by addition of guanidine hydrochloride (GuHCl). This is ascribed that the peptide tightly forms a helical structure in TFE, whereas it is stretched by GuHCl. FRET property strongly depends on the distance between fluorescein and rhodamine B, and hence, these FRET behaviors were clearly observed. On the other hand, no significant FRET change of 3mer peptide (AYC) was observed. This is probably due to no distance change, although CD spectrum of AYC altered. Subsequently a FRET-based sensor peptide for a specific peptide was designed. A 10mer

© 2002 American Chemical Society

sensor peptide (GSYEADRGGC) with a specific affinity to the 12mer peptides was modified with rhodamine B and fluorescein at both terminals, respectively. FRET property was investigated before and after forming a complex with the 12mer peptides as analyte molecules. The resulting fluorescence spectra clearly changed by addition of the analyte peptides. The FRET from fluorescein to rhodamine B was reduced by forming a complex. This is probably due to conformational change and quenching by addition of the analyte peptide. The FRET behavior strongly depended on sequence of the analyte peptides as well as analyte concentration. Furthermore, dissociation constants between the sensor and analyte peptides were estimated to be around 2 x 10^{-5} M.

INTRODUCTION

Biomacromolecules such as proteins have been employed as molecular recognition materials of biosensors (1-3). However, proteins are often denatured by heat, acid, alkali, or organic solvent. In order to avoid denaturation, biosensing should be carried out under mild condition. It has also been difficult to measure synthetic compounds because there is no enzyme or antibody corresponding to them. To solve these problems, we focused synthetic peptides that are much smaller than a protein. Molecular recognition ability of peptide is not inferior to protein, since an antibody recognizes antigen by a small peptide region called hypervariable loop or complementarity determining region. A small peptide has no catalytic activity but forms a complex with a particular target substrate, and the peptide conformation will change. In case the conformational change can be detected at high sensitivity, the peptide will be a sensor molecule for the target substance.

Fluorometry is one of the best candidates for highly sensitive detection method. A FRET method has often been used for detection of molecularly conformational change (4-7). If a donor fluorescent dye locates close to acceptor in the same molecule, FRET from the donor to the acceptor is observed. FRET behavior strongly depends on the distance between the donor and acceptor. Therefore, conformational change due to complex formation or cleavage can be detected by measuring fluorescence behavior.

In this work, sensor peptides based on FRET were designed. Prior to developing a sensor peptide, a FRET-based indicator peptide for the secondary structure was designed and synthesized. Principle of the indicator peptide is illustrated in Figure 1. Helical peptides, AY(KAAAA)$_n$KAC (n = 1, 2, 3), were selected as the model indicator molecules, and the N-terminal α-amino group and C-terminal cysteine side-chain were modified with rhodamine and fluorescein, respectively. Fluorescence property of these peptides were investigated.

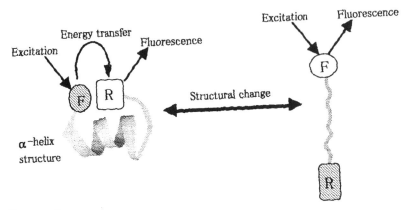

Figure 1 Principle of FRET-based indicator peptide for the secondary structure

A FRET-based sensor peptide for a specific peptide was designed. A 10mer sensor peptide (GSYEADRGGC) with a specific affinity to the 12mer peptides, which were screened from phage display random peptide library, was modified with rhodamine B and fluorescein at both terminals, respectively. Principle of the measurement is shown in Figure 2. The FRET property was investigated before and after forming a complex with the 12mer peptides as analyte molecules.

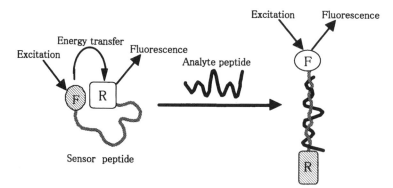

Figure 2 Principle of FRET-based sensor peptide

EXPERIMENTAL

Chemicals

Pre-packed N-α-9-fluorenylmethoxycarbonyl (Fmoc)-protected amino acids containing benzotriazole-1-yl-oxy-tris-pyrroridino-phosphonium hexafluorophosphate (PyBOP), which activates carboxyl group, were purchased from Shimadzu. The following side-chain protecting groups were chosen; 2,2,5,7-pentamethylchromane-6-sulphonyl (Pmc) for Arg, trityl (Trt) for Asn, Cys, Gln, and His, t-butyl ester ($OtBu$) for Asp and Glu, t-butoxycarbonyl (Boc) for Lys, and tBu for Ser, Thr and Tyr. Fmoc amino acids without side-chain protecting group were used for the other amino acids. A Cys(Trt)-preloaded Wang resin (Shimadzu) was employed. All deprotection and cleavage can be achieved using trifluoroacetic acid (TFA). N-Hydroxybenzotriazole (HOBt) and N-methyl morpholine (NMM), which were used for amide coupling reaction, were purchased from Shimadzu. Dimethyl formamide, t-butylmethylether, TFA, thioanisole, anhydrous diethyl ether, 1,2-ethanedithiol were from Nacalai Tesque, piperidine for deprotecting Fmoc-amino group was from Wako Pure Chemicals. Ethylmethyl sulfide, thiophenol and 2-methyl indole were from Tokyo Chemical Co.

Rhodamine B sulfonyl chloride and fluorescein-5-maleimide were from Molecular Probe. Trifluoroethanol (TFE) and guanidine hydrochloride (GuHCl) were from Wako.

Synthesis of FRET-based conformational indicator peptides

Each peptide with a different length includes KAAAA sequence which forms a α-helical structure. Sequences are listed as follows; Peptide 1F: Rho-AYC(Flu), 2F: Rho-AYKAAAAKAC(Flu), 3F: Rho-AY(KAAAA)$_2$KAC(Flu), 4F: Rho-AY(KAAAA)$_3$KAC(Flu), where Rho- and (Flu) stand for N-α-rhodamine B and side-chained fluorescein, respectively. The peptides were synthesized with an automated peptide synthesizer (Shimadzu PSSM-8) using a standard Fmoc protocol. Rhodamine B sulfonyl chloride was conjugated to the N-α-amino terminus, while the peptides were still attached to the resin and side-chain groups were still protected. Then, the peptides were cleaved from the resin and deprotected in a mixture of 90% TFA, 5% thioanisole, 5% 1,2-ethanedithiol for 2 hours at room temperature, and precipitated by cold diethyl ether, and lyophilized. The resulting peptides were allowed to react with fluorescein maleimide which is conjugated to cysteine residue. Synthesis of the FRET-based conformation indicator peptides is summarized in Figure 3.

The doubly labeled peptides were purified by reverse-phase HPLC system (Tosoh), eluted isocratically. Reverse phase octadecyl column (Tosoh ODS-80TS), UV-visible and fluorescence detectors (Tosoh UV-8020, FS-8020) were used. Eluent was 0.1% TFA-containing aqueous solution/acetonitrile (55:45). A fraction indicating fluorescence due to both fluorescein and rhodamine B was collected.

252

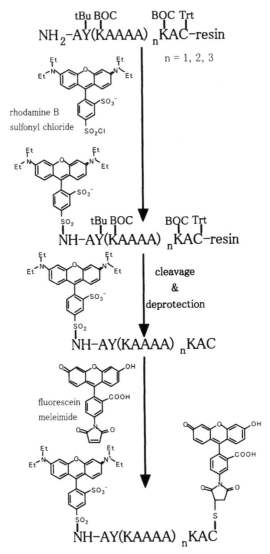

Figure 3 Synthesis of the FRET-based conformational indicator peptides

Matrix-assisted laser desorption ionization time of flight (MALDI-TOF) mass spectrometric analysis (Applied Biosystems Voyager RP) was carried out for all the peptide without fluorescence dyes. α-Cyano-4-hydroxycinnamic acid was purchased from Aldrich and used for matrix substance of MALDI.

Synthesis of FRET-based sensor and analyte peptides

The sequence of the sensor peptide and analyte peptides had already obtained in our group using phage display random peptide library. The sensor peptide, Rho-GSYEADRGGC(Flu), was synthesized by the same protocol as the FRET-based indicator peptides. Several kinds of 12mer peptides as analytes were also synthesized with a standard Fmoc chemistry. A 90% TFA solution including 5% thioanisole, 5% 1,2-ethandithiol and additional 1 mg/mL 2-methylindole was used as a cleavage cocktail for a peptide including Trp. A cleavage cocktail for the peptide containing Arg was composed of 82.5% TFA, 3% ethylmethyl sulfide, 5% water, 5% thioanisole, 2.5% 1,2-ethanedithiol and 2% thiophenol. Deprotection and cleavage reaction for Arg-containing peptides was allowed for 8 hours. The FRET-based sensor peptide was purified with reverse-phase HPLC in the same manner as the indicator peptides.

Spectrophotometry of fluorescent peptides

Fluorescence measurement of the fluorescent peptides was performed with a fluorescence spectrophotometer (Jusco FP-777). Excitation wavelength was fixed at 492 nm for fluorescein, and emission spectrum was measured. Each concentration of FRET-based indicator peptides was 5 μM in 0.1 M citric acid buffer (pH 3), citric acid buffer +15% TFE, citric acid buffer +3M GuHCl. Circular dichroism (CD) spectra were recorded with Jusco J-720, and concentration of the peptides was 80 μM for CD measurement.

Each concentration of FRET-based sensor peptide was 5 μM in 0.2 M phosphate buffer (pH 7), and the concentration of the analyte peptides was varied at 5, 10, 20, 50, and 100 μM.

RESULTS AND DISCUSSION

FRET-based conformational indicator peptides

Figure 4 shows fluorescence spectra of peptide 1F, 2F, 3F and 4F. Every fluorescence spectrum demonstrates two peaks at 520 nm and 580 nm corresponding to the emission from fluorescein and rhodamine B, respectively. This indicates that FRET occurred from fluorescein to rhodamine B, since rhodamine B is hardly excited at this wavelength, 492 nm. In Figure 4b, 4c and 4d, TFE addition to the citrate buffer solution containing the fluorescent peptides caused increase in fluorescence intensity from rhodamine B, in contrast with decrease in fluorescence intensity from fluorescein. In case GuHCl was added to the solution, the emission intensity from rhodamine B decreased.

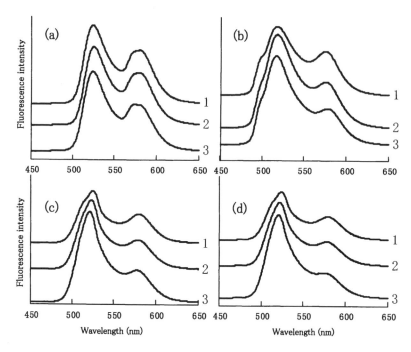

Figure 4 Fluorescence spectra of peptide. (a)1F(Rho-AYC(Flu)), (b) 2F (Rho-AYKAAAAKAC(Flu)), (c)3F (Rho-AY(KAAAA)₂KAC(Flu)), (d) 4F (Rho-AY(KAAAA)₃KAC(Flu)). Each peptide (5μM) was dissolved in 0.1 M citrate buffer (1) + 15% TFE, (2) no additive, (3) + 3M GuHCl. Excitation wavelength was fixed at 492 nm for fluorescein.

Table I   FRET index I_R/I_F of conformation indicator peptides

Peptide No.	1F (3)	2F (10)	3F (15)	4F (20)
peptide + TFE	0.68	0.64	0.56	0.49
peptide only	0.69	0.49	0.43	0.38
peptide + GuHCl	0.66	0.40	0.35	0.26

Numbers of amino acid residue are indicated in parenthesis.

The FRET property of these indicator peptides was evaluated quantitatively. Table I shows FRET index that was given by fluorescence intensity ratio of rhodamine B to fluorescein, i.e., I_R/I_F. In this table, FRET index increased with addition of TFE and decreased with GuHCl for peptide 2F, 3F and 4F. This is due to the distance change between rhodamine B and fluorescein. The peptides form tight structure with TFE and were stretched by GuHCl. However, No significant fluorescence change was observed for peptide 1F. This is due to no distance change between rhodamine B and fluorescein. Moreover, comparing among peptide 2F, 3F and 4F, FRET index decreased in this order. This is ascribed that the peptide 4F is the longest molecule of the three, and hence, FRET intensity from fluorescein to rhodamine showed the lowest value.

CD analysis was carried out for the peptide without fluorescent dye (Figure 5). The CD spectra of peptide 2, 3 and 4 in the citrate buffer showed negative maximum at 208 and 222 nm that are characteristic of a helical peptide. TFE addition to the solution caused increase of negative maximum. This indicates that the content of helical structure increased. In case GuHCl was added to the solution, negative maximum disappeared. This indicates that the peptides no longer formed helical structure. Therefore, we can conclude that conformational change of the peptide caused change of the FRET property.

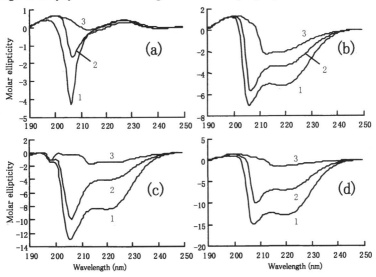

Figure 5   CD spectra of (a) peptide 1 (AYC), (b) peptide 2 (AYKAAAAKAC), (c) peptide 3 (AY(KAAAA)₂KAC), (d) peptide 4 (AY(KAAAA)₃KAC). Each peptide (80 μM) was dissolved in 0.1 M citrate buffer (1) + 15% TFE, (2) no additive, (3) + 3M GuHCl. Molar ellipticity in 10^4 deg cm² dmol⁻¹ is revealed.

No change occurred in the fluorescence spectra of 1F in spite of addition of TFE or GuHCl, whereas the CD spectra of peptide 1, AYC, showed a conformation change. Probably, conformation change of peptide 1F occurred, but no distance change between rhodamine B and fluorescein occurred.

FRET-based sensor peptide

Figure 6 shows fluorescence spectra of the sensor peptide SP, Rho-GSYEADRGGC(Flu), in the absence and presence of the model analyte Peptide A, HPMNMHRHGYIY. The fluorescence spectrum of SP indicates two peaks at 520 nm and 580 nm corresponding to conjugated fluorescein and rhodamine B. Addition of peptide A, obtained from phage-based screening, to the solution initiated decreasing the fluorescence emission from rhodamine B, compared with intensity from fluorescein. This probably indicates that the conformation of SP changed to be unfold than before addition of peptide A. Otherwise the energy transfer from fluorescein to rhodamine B was prevented by the complex formation, and quenching occurred. Since not only fluorescence intensity of rhodamine B but also that of fluorescein was reduced, it is difficult to analyze the details of these results, and hence, use of the FRET index is suitable for evaluation.

Other model analyte peptides resulting from phage display screening were also examined. The FRET index I_R/I_F of SP before and after addition of analyte was evaluated. Figure 7 shows the FRET index of SP as a function of analyte peptide concentration. Peptide A, B and C obtained from phage display library include homologous amino acid sequences, and the FRET index change by peptide A was observed to be as great as peptide B. Furthermore, peptide D is homologous to peptide E, and the FRET index change by these two peptides indicated the same level. No significant change was observed for addition of peptide R with random amino acid sequence. Hence, this FRET-based sensor peptide was selective to the specific peptides.

Dissociation constant K_D value between SP and analyte peptide was estimated using FRET index changes. The reciprocal value of the ratio of SP forming the complex was plotted against the reciprocal of analyte peptide concentration. When analyte peptide concentration is much greater than the complex concentration, the slope of the double reciprocal plot approximately corresponds the K_D value. The K_D values for peptide A and B were determined to be around 2×10^{-5} M.

In this research, we designed and synthesized FRTE-based sensor peptides. In case peptides are consist of 10 to 20 amino acid residues, the FRET property change was clearly observed, and the conformational change can be detected at high sensitivity. When the analyte peptide was added to the FRTE-based sensor peptide solution, the FRET property change was observed. The details of the mechanism for fluorescence property change is unclear, but it is possible to detect the analyte peptide by measuring FRET property of the sensor peptide.

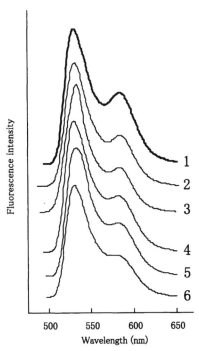

Figure 6 Fluorescence spectra of sensor peptide. Fluorescence spectra were measured for 5 µM sensor peptide in the (1) absence, and presence of (2) 5, (3) 10, (4)20, (5) 50, (6) 100 µM peptide A. Excitation wavelength was fixed at 492 nm.

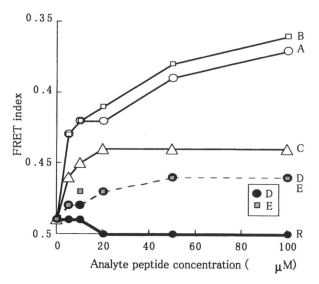

Figure 7 FRET index of sensor peptide as a function of analyte peptide concentration. The FRET index was defined as I_R/I_F. The following analyte peptide sequences were used; A: HPMNMHRHGYIY, B: TPVINYMSGSHF, C: QIPLMKGPGYMY, D: HPPMDFHKAMTR, E: DPPTVLPKLAYR, R: NADLPSFESNGC.

REFERENCES

1. *Biosensor Principle and Applications*; Blum, L. B.; Coulet, P. R., Ed.; Marcel Dekker: New York, NY, 1991.

2. Koide, S.; Yokoyama, K. *J. Electroanal. Chem.* **1999**, *468*, 193-201.

3. Yokoyama, K.; Ikebukuro, K.; Tamiya, E.; Karube, I.; Ichiki, N.; Arikawa, Y. *Anal. Chim. Acta* **1995**, *304*, 139-145.

4. Adams, S. R.; Harootunian, A. T.; Buechler, Y. J.; Taylor, S. S.; Tsien, R. Y. *Nature* **1991**, *349*, 694-697.

5. Godwin, H. A.; Berg, J. M. *J. Am. Chem. Soc.* **1996**, *118*, 6514-6515.

6. Tyagi, S.; Krammer, F. R. *Nature Biotechnol.* **1996**, *14*, 303-308.

7. Miyawaki, A.; Llopis, J.; Heim. R.; McCaffery, J. M.; Adams, J. A.; Ikura, M.; Tsien, R. Y. *Nature* **1997**, *388*, 882-887.

Chapter 21

Immobilized Fluorescent Liposome Column for Bioanalysis and Signal Amplification

Xue-Ying Liu, Chikashi Nakamura, and Jun Miyake*

Tissue Engineering Research Center, National Institute for Advanced Industrial Science and Technology (AIST), 1–1–1 Higashi, Tsukuba, Ibaraki 305–8562, Japan

Fluorescent dye, calcein-entrapped unilamellar liposomes were stably immobilized in gel beads by avidin-biotin binding to construct the stationary phase for immobilized liposome chromatography (ILC). High-immobilized amounts, excellent stability and membrane integrity of the immobilized liposomes provide the basis for their applications in bioanalysis and biotechnology. This chapter reports the following subjects using the immobilized fluorescent liposome column: (1) Sensitively analysis interactions between the drugs, lipophilic cations, peptides and proteins with lipid bilayers. Retardation of solutes was monitored using a UV detector; meanwhile perturbation of the membranes by some solutes gave rise to the calcein leakage monitored by an online fluorescent detector. (2) Rapidly detection of Phospholipase A_2 (PLA_2)-catalyzed membrane leakage. (3) Detection of small molecules using a competitive immuno-reaction and sensitization by PLA_2-catalyzed dye leakage from immobilized liposomes.

Liposomes have two distinct domains of lipid bilayer membrane and interior aqueous cavity. Therefore, they have been used as model membranes for the study of interactions between proteins, peptides, drugs and other biologically important substances with cell membranes (1-4), and as sensitive probes for

© 2002 American Chemical Society

signal amplification inherent to liposomal encapsulation of different indicator molecules (5, 6). For chromatographic studies, liposomes, proteoliposomes and membrane vesicles have been immobilized in solid supports consisting of gel beads for the quantitative analysis of drug- or peptide-membrane interactions using immobilized liposome chromatography (ILC) (7-9). In order to construct a homogeneous lipid membrane stationary phase, we successfully immobilized the unilamellar liposomes in gel beads by avidin-biotin specific binding (10), which stabilized immobilized liposomes greatly and prolonged the column lifetime over one year. Variations in retention volume depend on the extent of solute-membrane interaction and can be precisely measured to membrane partitioning coefficient (11-14). However, it is often difficult to detect weak interactions of proteins and peptides with membranes from retention volume using ILC analysis. Liposomes containing entrapped fluorescent labels, such as, calcein, within the aqueous space can be used to investigate the changes in membrane permeability (15, 16). This implies that fluorescent dye-entrapped liposome chromatography may provide information of both solute retardation and membrane permeability by applying solutes or membrane destabling agents to the liposome column. The present chapter describes the applications of the fluorescent liposome column in analysis the weak interaction (17), detection of PLA_2-catalyzed membrane leakage (18), which was applied to the detection of PCB using a competitive immuno-reaction (19).

Preparation of calcein-entrapped liposome column

Avidin was coupled to chloroformate activated Sephacryl S-1000 gel or TSK G6000PW gel (denoted hereafter as Sephacryl and TSK, respectively) at a concentration of 3 mg protein/ml gel as described in Ref. 10. Alternatively, avidin was coupled to CNBr-activated Sepharose 4B at 3.0–3.5 mg/ml of gel beads according to the manufacturer's specifications. The avidin-gels were stored at 4°C in buffer H (10 mM HEPES, 150 mm NaCl, 0.1 mM Na_2-EDTA, pH 7.5) supplemented with 3 mM NaN_3 until used (10). EPC supplemented with 2 mol% of biotin-cPE was evaporated using a rotary evaporator to form a dry film. The film was flashed with nitrogen, and kept under high vacuum for at least 3 h, and then dispersed in 100 mM calcein solution (pH 7.5) to form MLVs. Small unilamellar liposomes and large unilamellar liposomes (denoted as SUVs and LUVs, respectively) were prepared by probe sonication and extrusion on two-stacked polycarbonate filters of 100-nm pore size, respectively, as described elsewhere (10). For immobilization (10), the calcein-entrapped biotinylated liposomes were gently mixed with moist avidin-gel of TSK, Sephacryl, or Sepharose by rotation for 2-3 h at 23°C or overnight at 4°C under nitrogen. Nonimmobilized liposomes and non-entrapped calcein were then removed by washing with buffer H on the 10-μm filter. The phospholipid amounts in aliquots of the immobilized gels or in eluates with 200 mM β-OG (octylglucoside) from the gel beds were determined as phosphorus, as described by Bartlett (*20*).

ILC-fluorescent analysis

The gel beads containing avidin-biotin immobilized liposomes were packed into a 5 mm I.D. × 2–5.5 cm gel bed in a glass column (HR 5/5, Pharmacia Biotech). Solutes such as pharmaceutical drugs, phosphonium cations, peptides, proteins and enzymes were applied to the immobilized-liposome gel bed and eluted with buffer H at a flow rate of 0.3 or 0.5 ml/min as described in Ref. (*17-19*). The HPLC system consisted a column oven (CO–8020, Tosoh) equipped with a sample injector, a HPLC pump (CCPM-II, Tosoh), a UV-detector (UV–8010, Tosoh), a Fluorescent-detector (FS-8020, Tosoh), and a recorder interfaced with an IBM computer.

The chromatographic runs were monitored using a UV-detector set at 220 nm for drugs and peptides, 267 nm for phosphonium cations and bovine carbonic anhydrase (CAB), and 280 nm for Phospholipase A_2 (PLA$_2$), respectively, a fluorescence detector at excitation and emission wavelength of 490 nm and 520 nm, respectively.

Retention of the solute was expressed as membrane partitioning coefficient, K_{LM} (*11-14*). Further, perturbation of the liposomal membranes by interactions resulting in increased permeability can be observed by the calcein released from the immobilized liposomes. The amount of released calcein can be estimated from the peak area of eluted calcein.

Characterization of the avidin-biotin-immobilized liposomes

Biotinylated liposomes with entrapped fluorescent dye, calcein, were immobilized in a variety of avidin-gel beads, and their characterization is summarized in Table 1. The LUVs were immobilized only in Sephacryl or TSK gel beads, because the pore sizes of these gels are up to 400 nm and 500 nm, respectively, favoring the immobilization of the large unilamellar liposomes (*10 and references therein*). The specific trapped volumes of the avidin-biotin immobilized SUVs and LUVs (Table 1) were similar to the literature data for the SUVs and LUVs, respectively (*21*). These values together with the mean size of the SUVs (31±13 nm) or LUVs (109±7 nm) indicate the unilamellarity of the avidin-biotin-immobilized liposomes. Only 0.4-1.1% of the initially entrapped calcein was released from the immobilized SUVs and LUVs during storage for

Table I. Characterization of immobilized liposomes in Hepes buffer

Type of liposomes	Mean diameter (nm)	Type of avidin-gels	Amount (μmol lipid/ml gel)	Trapped volume (μl H$_2$O/ μmol lipid)	Leakage (%)
SUVs	31 ± 13	Sepharose	40.0 ± 4.9	0.54 ± 0.07	0.53 ± 0.11
		Sephacryl	33.2 ± 3.8	0.68 ± 0.04	0.77 ±0.12
		TSK	33.6 ± 1.2	0.60 ± 0.04	0.74 ± 0.21
LUVs	109 ± 7	Sephacryl	40.9 ± 3.7	2.30 ± 0.03	0.43 ± 0.05
		TSK	36.9 ± 3.3	1.90 ± 0.04	0.47 ± 0.12

NOTE: The mean value (n=4) is given.

SOURCE: Reproduced from Reference 18. Copyright 2001 Academic Press.

one week at 4°C, indicating that the avidin-biotin-immobilized liposomes retained their integrity well. Several avidin molecules coupled to the gel may offer multiple-site binding per each biotinylated liposome, as illustrated in Fig. 1. The avidin-biotin site binding did not perturb the liposomal membrane, as it was shown that the biotinylated liposomes during streptavidin- mediated aggregation were likely to be spherical and unstressed (22). In contrast to the free-form liposomes, which were largely aggregated and precipitated upon storage, the liposomes immobilized in gel beads are probably protected from such

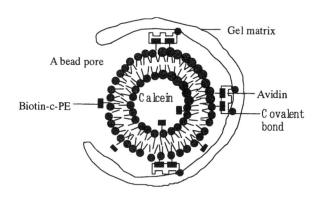

Figure 1. Illustration of a liposome immobilized by avidin-biotin in a gel matrix. (Reproduced from Reference 10. Copyright 1998 Elsevier Science.)

aggregation by the gel matrix. The immobilized liposomes showed excellent stability during long-term storage and upon several chromatographic runs (*10, 11, 13, 17*). The loss of lipid and entrapped calcein from the gels was 3-5% after 5 months storage at 4°C under nitrogen (*10, 17*). Stability in the flow chromatographic liposome column was demonstrated by the fact that only 7% of the liposomes was lost after 60 or 75 runs at different temperatures (*10, 13*).

Analysis of solute-membrane interaction

The fluorescent liposome chromatography was successfully used as a sensitive method to detect solute-membrane interactions; in some cases, this method can detect an interaction, which is too weak to be detected by the chromatographic retardation described above where the retardation was monitored using a UV detector. Perturbation of the membranes by some solutes gives rise to the leakage of the entrapped calcein, which can be detected by an on-line flow-fluorescent detector. Generally, fluorescent is more sensitive than UV.

Interactions of lipophilic cations or pharmaceutical drugs with membranes have been studied by immobilized liposome chromatography (*11, 17*). It has been shown that similar K_{LM} values of lipophilic cations or drugs were obtained by performing the ILC runs on two gel beds of different size (*11*) or two kinds of gel bed (*17*) containing different amounts of liposome. This is important from the practical point of view in the use of the ILC method for quantitative analysis of solute-membrane partitioning in the laboratory, since it is hardly possible to prepare a constant amount of immobilized liposomes in gel beds of the same dimension by batchwise procedures. It also meant that the entrapped calcein molecule has no effect on lipophilic cations partitioning into membranes. Nevertheless calcein leakage caused by lipophilic cations partitioning can be observed. Perturbation of the liposomal membranes by the binding of the lipophilic cations was observed from the leakage of calcein trapped in the immobilized liposomes (Fig. 2). The perturbation became more pronounced when the hydrophobicity of the solutes (i.e., the chromatographic retardation) was increased. The linear relationship between the solute hydrophobicity and the calcein leakage indicates that the lipophilic cations penetrate into the immobilized liposomal membranes. A similar relationship was found in the interactions between drug and liposomes by ILC (*17*). Further, the hydrophobic interaction between lipid acyl chain and hydrophobic residues of the synthesis peptide and the electrostatic interactions between the polar residues of the polylysine peptide and the phospholipid head groups can be revealed by the membrane leakage on the immobilized fluorescent liposome column, although the interactions were difficult to detect by their retardations (*17*). On the other

264

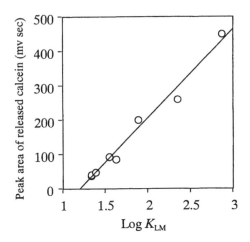

Figure 2. Correlation of Log K_{LM} and membrane leakage by lipophilic cations.
(Reproduced from Reference 11. Copyright 1999 Elsevier Science.)

hand, for the 0.5-2.5 M guaninium hydrochloride (GuHCl) denatured CAB, significant dye leakage from the liposomes was observed although the retardation was hard to measure (Fig. 3). It should be noted that there is no calcein leakage when only a 2.5 M GuHCl solution was applied to the fluorescent liposome column. Therefore, the membrane perturbation by CAB and retardation of CAB on ILC are thought to be due to protein-liposome interactions during the refolding process. Weak protein-membrane interactions can thus be revealed from the significant leakage of calcein from the liposomes. This provides additional useful information on the properties of the denatured proteins.

Detection of Phospholipase A_2–catalyzed membrane leakage

Phospholipase A_2 (PLA_2)-catalyzed membrane leakage has been largely studied using liposomes with an entrapped fluorescent probe *(23, 24)*. We employed ILC to study the PLA_2-catalyzed membrane leakage in order to investigate the hydrolytic action of PLA_2. By passing a small amount of PLA_2 through the liposome column, the release of entrapped calcein from the liposomes composed of substrate phosphatidylcholine could be monitored by the

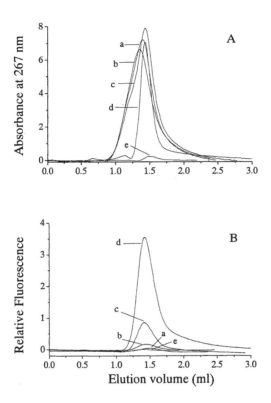

Figure 3. Chromatographic retardation (A) and membrane leakage (B) for CAB on the calcein-entrapping SUV-Sepharose column. The protein was preequilibrated in the presence of 0, 0.5, 1.5, 2.5 M GuHCl (curves a-d) before injection. Curve e represented only 2.5 M GuHCl applied to the liposome column. (Reproduced from Reference 17. Copyright 2001 Elsevier.)

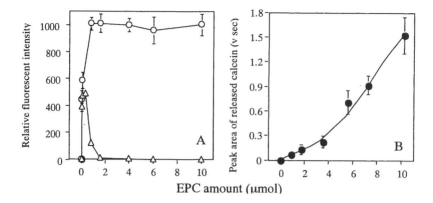

Figure 4. Comparison of PLA₂-catalyzed membrane leakage in free liposome suspensions (A) and immobilized liposome column (B). (A) After incubation of different amount of EPC LUVs with 0.4 μg PLA₂ for 1 h, the 2-ml mixtures were measured by a fluorescent spectrophotometer before (△)and after dilution (○). (B) The same amount of PLA₂ was applied to the liposome column with different immobilized amounts of EPC. The error bars represent standard deviation for three independent measurements. (Reproduced from Reference 18. Copyright 2001 Academic Press.)

ILC method using an on-line flow fluorescent detector. The PLA₂-catalyzed membrane leakage on the immobilized liposomes as studied by ILC was found to be affected by the gel pore size used for the immobilization, liposome size and as expected, by the concentration of calcium, but unaffected by the flow rate (*18*). We found that the largest PLA₂-induced calcein release from the liposome column was detected on LUVs immobilized on TSK or Sephacryl gel in the presence of 1 mm Ca^{2+} in the aqueous mobile phase. Figure 4 shows the PLA₂-catalyzed membrane leakage detected in the liposome free suspensions by batchwise fluorescent measurement and in immobilized liposomes by ILC-fluorescent assay for various EPC amounts. At high EPC amounts, the fluorescent signal of the released calcein in the free liposomes by PLA₂ was not detected due to the self-quenching, hence the dilution needed to be assayed. In addition, the fluorescence reached a maximal level in a low free liposome concentration. On the other hand, the fluorescent signal that originated from the leakage can be simply detected and significantly enhanced even by a higher amount of immobilized liposomes (Fig. 4). This may have potential applications in biosensors for signal amplification of the PLA₂-ligand complex as described below.

A novel biosensor for PCB detection

The polychlorinated biphenyls (PCBs) are industrial compounds or by-products that have been widely identified as environmental contaminants that have acute toxicity, carcinogenicity and binding activity to the cytosolic aramatic or aryl hydrocarbon receptor. The chemical and physical properties of PCBs make analysis difficult. Large amounts of these compounds remain in the environment, in use or in waste. Therefore detection and remediation of PCBs as well as other environmental pollutants (e.g. dioxin) become most important issue today. we developed a rapid and simple two-column detection system for environment pollutant chemicals. The novel method combined competitive immunoassay with ILC-fluorescent analysis was developed to detect the environmental contaminants, such as PCB. As described above, the liposome column can be stored over one year (17) and the fluorescent dye leakage from liposomes caused by PLA_2 can be more sensitively and rapidly detected by ILC runs (18). These techniques allow us to extend the liposome chromatography technique into the field of environmental detection of chlorinated organics.

Two column detection system

For the competitive immuno-reaction, the anti-PCB antibodies were coupled to HiTrap Protein G column as the first column (19). The immobilized calcein-entrapped liposome gel beads were packed in an open column as the second column. Basis on that the competitive conjugate should resemble the analyte as closely possible (25), we selected the competitor of 3,4-dichloroaniline to mimic half domain of the PCB molecule. The PLA_2 conjugate was prepared by coupling the amino group of 3,4-dichloroaniline to the carboxyl group of PLA_2 activated by NHS (26). The PLA_2 conjugates remained PLA_2 catalytic activity, about half of native PLA_2.

PCB sample was dissolved in acetone to 100 µg/ml as a stock solution. Serial dilution using HEPES buffer was done to prepare PCB concentration range from 0.01-10 µg/ml. The mixtures of PLA_2 conjugate (100 µl, 10 µg/ml) with various PCB dilutions (100 µl, 0.01-10 µg/ml) were applied to the first column. After incubation at room temperature for at least 15 min, the unbound PLA_2 conjugate was eluted out using HEPES buffer. The eluate was injected to the second column. The fluorescent liposome column was eluted with HEPES buffer. The each 2 ml of collection was measured its fluorescent intensity. The antibody column was regenerated using acetate buffer (pH 4.0) followed by HEPES buffer, meanwhile the liposome column was washed with HEPES buffer to achieve a stable base line without fluorescent.

PCB detection by two-column using competitive immuno-reaction and sensitization by dye leakage from liposomes

Since the detection limitation of PCB was 1 µg/ml in the reference report (27), we studied the optimum concentrations of PLA$_2$ conjugate by applying the mixture of 1 µg/ml PCB and various concentrations of PLA$_2$ conjugate to the two-column PCB detection system. It was found that 10 µg/ml of PLA$_2$ conjugate caused higher effective fluorescent signal in the PCB detection (19).

As shown in Fig. 5, the calcein leakage from the liposomes caused by PLA$_2$ conjugate increased with the PCB concentration. There is no fluorescence increase when only PCB sample applied to the columns (open circles). It suggests that PCB itself does not affect on the membrane leakage. About 25 relative fluorescent intensity is noise level in the retardation of the sample solution in this detection system. A clear correlation between the fluorescent intensity and the concentration of PCB is observed in the 0.01–10 µg/ml region Concentration of PCB sample as low as 10 ppb was measurable using this system. The detection limit was 100 times lower than that reported by Charles et al. (27), using a conjugate of tetrachlorophenol and fluorescent dye, Cy5.29 for

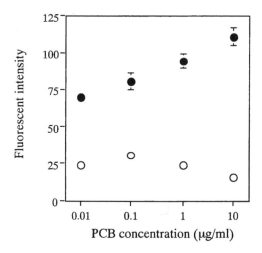

Figure 5. Calibration graph for PCB detection. ●, with PLA$_2$ conjugate; ○, without PLA$_2$ conjugate. The error bars represent standard deviation for three independent measurements. (Reproduced from Reference 19)

competitive binding. The PCB detection could be reproducibly detected using this biosensor system. The coefficients of variation in the measurement were under 7% (Fig. 5), and it takes less than 30 min for the whole detection process.

Concluding remarks

Stable and high-yield immobilization of the fluorescent dye-entrapped liposomes in the gel beads can be achieved by avidin-biotin multiple-site binding. The ILC method is easy, sensitive, and reproducible for estimating the interaction of a solute with a liposomal membrane even though the interaction is weak. Using the fluorescent liposome column, the interaction of solutes with membranes can be analyzed not only from their retardation on the column but also from the leakage of the liposomal membrane. PLA_2-catalyzed hydrolysis of phosphatidylcholine can be straightforwardly detected in the fluorescent leakage from the gel-bead-immobilized liposomes using ILC method. Compared with free liposome suspensions, the signal of fluorescent leakage from liposomes caused by PLA_2 hydrolysis are more sensitively and rapidly detected in the fluorescent liposome column by an ILC run. Therefore, a new sensitive method for thedetermination of small molecules was developed using a competitive immuno-reaction sensitized by PLA_2-catalyzed membrane leakage. Moreover, most of the entrapped calcein was retained in the immobilized liposomes, and the immobilized liposomes remain stable during the chromatographic runs and long-term storage. It was again demonstrated that the avidin-biotin immobilized liposomes have excellent stability (*11, 13, 17, 18*) and membrane integrity when used as a membrane stationary phase for analysis of solute-membrane interaction.

Acknowledgements

We would especially like to thank Dr. Q. Yang (Viral and Rickettsial Disease Department, Naval Medical Research Center) for his contribution to the ILC analysis of solute-membrane interactions. This work was supported by the Biomolecular Mechanism and Design Project of National Institute for Advanced Interdisciplinary Research, Agency of Industrial Science and Technology (AIST), The Ministry of International Trade and Industry (Japan).

References

1. Bangham, A. D.; Hill, M. W.; Miller, N. G. A. *Methods Membr. Biol.* **1974**, *1*, 1-68.
2. Rogers, J. A.; Choi, Y. W. *Pharm. Res.* **1993**, *10*, 913-917.

270

3. Nishiya, T.; Hui, L. C. *J. Biochem.* **1991**, *110*, 732-736.
4. Akhtar, S.; Basu, S.; Wickstrom, E.; Juliano, R. L. *Nucleic Acids Res.* **1991**, *19*, 5551-5559.
5. Roberts, M. A.; Durst, R. A. *Anal. Chem.* **1995**, *67*, 482-491.
6. Pinnaduwage, P.; Huang, L. *Clin. Chem.* **1988**, *34*, 268-272.
7. Yang, Q.; Lundahl, P. *Biochemistry* **1995**, *34*, 7289-7249.
8. Zhang, Y.; Aimoto, S.; Lu, L.; Yang, Q.; Lundahl, P. *Anal. Biochem.* **1995**, *229*, 291-298.
9. Lundahl, P.; Beigi, F. *Adv. Drug Deliv. Rev.* **1997**, *23*, 221-227.
10. Yang, Q.; Liu, X.-Y.; Ajiki, S.-I; Hara, M.; Lundahl, P.; Miyake, J. *J. Chromatogr. B.* **1998**, *707*, 131-141.
11. Yang, Q.; Liu, X.-Y.; Umetani, K.; Kamo, N.; Miyake, J. *Biochim. Biophys. Acta* **1999**, *1417*, 122-130.
12. Yang, Q.; Liu, X.-Y.; Umetani, K.; Ikehara, T.; Miyauchi, S.; Kamo, N.; Jin, T.; Miyake, J. *J. Phys. Chem. B* **2000**, *104*, 7528-7534.
13. Liu, X.-Y.; Yang, Q.; Kamo, N.; Miyake, J. *J. Chromatogr. A* **2001**, *913*, 123-131.
14. Liu, X.-Y.; Yang, Q.; Hara, M.; Nakamura, C.; Miyake, J. *Material Science Engineering: C* **2001**, *17*, 119-126.
15. New, R. R. C. *Liposomes: A Practical Approach;* IRL Press: Oxford, 1990; pp128-134.
16. Allen, T. M. *Calcein as a toll in liposomes methodology;* Gregoriadis, G., Ed.; CRC Press, Boca Raton, FL, 1984; Vol. III, p 177.
17. Liu, X.-Y.; Yang, Q.; Nakamura, C.; Miyake, J. *J. Chromatogr. B* **2001**, *750*, 51-60.
18. Liu, X.-Y.; Nakamura, C.; Yang, Q.; Miyake, J. *Anal. Biochem.* **2001**, *293*, 251-257.
19. Liu, X.-Y.; Ph. D. thesis, Graduate School of Pharmaceutical Science, Hokkaido University, Sapporo, 2001.
20. Bartlett, G. R. *J. Biol. Chem.* **1959**, *234*, 466-468.
21. Hope, M. J.; Bally, M. B.; Mayer, L. D.; Janoff, A. S.; Gullis, P. R. *Chem. Phy. Lipids* **1986**, *40*, 89-107.
22. Chiruvolu, S.; Walker, S.; Israelachvili, J.; Schmitt, F.-J.; Leckband, D.; Zasadzinski, J. A. *Science* **1994**, *264*, 1753-1756.
23. Fugman, D. A.; Shirai, K. R.; Jackson, L.; Johnson, J. D. *Biochim. Biophys. Acta* **1984**, *795*, 191-195.
24. Shimohigashi, Y.; Tani, A.; Matsumoto, H.; Nakashima, K.; Yamaguchi, Y.; Oda, N.; Takano, Y.; Kamiya, H.; Kishino, J.; Arita, H.; Ohno, M. *J. Biochem.* **1995**, *118*, 1037-1044.
25. Jung, F.; Gee, S.; Harrison, R.; Gooddrow, M.; Karu, A.; Braun, A.; Li, Q.; Hammock, B. *Pestic. Sci.* **1989**, *26*, 303-317.
26. Sehgal, D.; Vijay, I. K. *Anal. Biochem.* **1994**, *218*, 87-91.
27. Charles, P. T.; Conradm, D. W.; Jacobs, M. S.; Bart, J. C.; Kusterbeck, A. W. *Bioconjugate Chem.* **1995**, *6*, 691-694.

Chapter 22

Analysis and Optimization of Biopanning Process of Phage Display Libraries

Yoshio Katakura, Guoqiang Zhuang, Satoshi Uchida,
Tetsuo Furuta, Takeshi Omasa, Michimasa Kishimoto,
and Ken-ichi Suga

Department of Biotechnology, Graduate School of Engineering, Osaka 565–
0871, Japan

A kinetic model describing the affinity selection process of phage display libraries was established and verified experimentally using ribonuclease A (RNase A) and anti-RNase A single chain Fv phage antibody. Desorption of target molecules from a solid phase and orientation of the epitopes of adsorbed target molecules are taken into account in this model. The ratio of the effective antigen to the total antigen was estimated to be $0.0127(+/-)0.0018$ when RNase A was immobilized on polystyrene beads by passive adsorption. The model can faithfully describe the recovery of the phage antibody in a round of biopanning based on the following parameters; the effective concentration of RNase A on the beads, the desorption rate constant of RNase A from the beads, the dissociation constant and dissociation rate constant of the phage antibody from RNase A, and the time for blocking, binding and washing in the biopanning process. The causes of the typical problems in biopanning are explained quantitatively by the model and the practical solutions for the problems are discussed.

© 2002 American Chemical Society

271

272

A phage display system is one of most powerful tools for isolating specific binders (1-3). Screening of phage libraries, referred to as "biopanning," analogous to the process of "panning for gold" is based on the affinity between a target of interest immobilized on a solid phase and a ligand displayed on a phage. Although, numerous specific binders have been isolated from phage display libraries, many researchers have often encountered the following problems.

· Poor reproducibility: The isolated clone against the same target is sometimes different even when the same researcher screens the same library by the same protocols.

· High background: Even after three or more rounds of biopanning, most of the clones in the output are negative.

· Low affinity: The isolated clones show a low affinity against their targets.

These problems are likely to be the result of unsuitable experimental conditions for biopanning. Until now, most biopanning protocols have been derived from that of Parmley and Smith (4). However, since their protocol is based on experience, most researchers have been foced to solve the problems by trial and error. The purpose of this study is to explain the causes of these problems and to show the practical solution for them based on the kinetic model describing the biopanning process (5).

Materials and Methods

Bovine pancreatic ribonuclease A (RNase A) purchased from Sigma (Code; R-5500, Tokyo) was dialyzed against distilled water prior to use. The gene encoding the anti-RNase A single chain Fv (scFv) 3A21 (6) was cloned into the phargimid vector pCANTAB5E (Amersham Pharmacia Biotech). The soluble anti-RNase scFv and the phage antibody were prepared by infecting Escherichia coli HB2151 and TG1, respectively, with the phage displaying anti-RNase A scFv according to the manufacturer's instruction. The culture supernatant was loaded onto an RNase A Sepharose 4B column (5) (bed volume 2 ml) and washed extensively with PBS (0.2 g of KH_2PO_4, 2.9 g of $Na_2HPO_4 \cdot 12H_2O$, 8.0 g of NaCl, and 0.1 g of KCl in one liter of deionized water, pH 7.2). The bound scFv or phage was eluted with 0.1 M glycine-HCl (pH 2.2). The eluent was neutralized by adding 1 M Tris-HCl (pH 8.0) and dialyzed against PBS. The concentration of the scFv solution was calculated based on the absorbance at 280 nm. The concentration of the phage antibody was titrated on E. coli TG1 cells as colony-forming units. Polystyrene paramagnetic microparticles (2.6% suspension, diameter 1–2 μm) were from Polysciences, Inc. (Warrington, France). The polystyrene paramagnetic beads (2.6 mg) were washed three times with 0.1 M sodium carbonate buffer (pH 9.6) and were then coated with RNase A (0.1 ml of 4.5 μM solution in the same buffer) at 4°C for 18 h with a gentle rotation. After washing the beads three times with PBS using a magnet, the beads were blocked by incubating in PBSM (PBS containing 2% Block Ace (Yukijirushi, Tokyo)) at 30°C with a gentle rotation. The beads were then washed three times with PBS and incubated in a solution of the purified anti-RNase A phage antibody (10^8 cfu in 0.2 ml PBSM) at 30°C for one hour with gentle rotation. The beads were then washed three times with PBS and incubated in 1 ml of PBSM at 30°C. After washing the beads three times with PBS, the phages bound to the beads were recovered by incubating the beads in 0.1 M triethanolamine and titrated on *E. coli* TG1 cells as colony-forming units.

Results and Discussion

A kinetic model for biopanning process

Mandecki *et al.* established the first model for a biopanning process (7), describing the relationship between the binding fraction of a phage antibody and the concentration of its antigen based on the Langmuir adsorption isotherm. That same year, Kretzschmar *et al.* introduced a simple relationship between the binding fraction and the antigen concentration based on mass-balance under a state of equilibrium (8). Based on these two studies, Levitan established a stochastic model and calculated the relative output of a clone assuming an affinity distribution in a library (9). In these studies, the binding fraction of a phage clone displaying an antibody was expressed as

$$F = \frac{C}{K_d + C} \exp(-k_{off} t_w) \tag{1}$$

where C is the total concentration of the antigen, K_d and k_{off} are the dissociation constant and the dissociation rate constant of the antibody from the antigen, respectively, t_w is the time required for the washing process. In these studies, the antigen concentration was identified as an important parameter affecting the results of biopanning. However, the concentration of the antigen during biopanning is not constant because adsorption of proteins on a solid phase is a partially reversible dynamic process (10, 11). After the coating process (incubation of a solid support in an antigen solution), a significant part of

274

adsorbed antigen molecule desorbs from the solid support during the blocking, binding and washing process (Figure 1, left). In addition, since access to a part of the epitope is hindered by the solid phase, blocking proteins or the antigen molecules themselves and since a part of the antigen molecules adsorbed on the solid phase are denatured by adsorption (*12–14*), only a part of the antigen molecules on the solid phase are effective for phage binding (Figure 1, right).

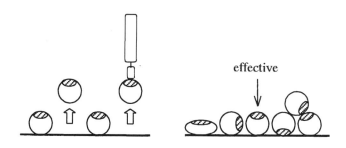

Figure 1 Desorption of antigen and orientation of the epitope.

Therefore, we established a kinetic model considering the antigen desorption and the effective antigen concentration (*5*). The binding fraction of a phage displaying an antibody is expressed as

$$F = \frac{\beta[\Gamma_0^{irr} + \Gamma_0^r \exp(-k_{des}t_{b+b+w})]\dfrac{S}{V}}{K_d + \beta[\Gamma_0^{irr} + \Gamma_0^r \exp(-k_{des}t_{b+b})]\dfrac{S}{V} + [\Gamma_0^r \exp(-k_{des}t_b) - \Gamma_0^r \exp(-k_{des}t_{b+b})]\dfrac{S}{V}} \exp(-k_{off}t_w)$$

(2)

where β is the ratio of the effective antigen to the total antigen on the solid phase, Γ_0^r and Γ_0^{irr} are the densities of the antigen reversibly and irreversibly adsorbed on the solid phase, respectively, k_{des} is the desorption rate constant of the antigen from the solid phase, S is the total surface area of the solid phase, V is the volume of the system, t is time (subscripts; b, time required for the blocking process; b+b, total time for the blocking and binding process; b+b+w, total time for the blocking, binding and washing process, w, time required for the washing process). The details of the introduction of Eq. (2) and the determination of the parameters are described in our previous study (*5*).

Determination of the Parameters and Verification of the Model

When the antigen (RNase A) was coated on the polystyrene bead by passive adsorption and the anti-RNase A phage antibody was panned against the coated beads, the following parameters in Eq. (2) were determined experimentally (see reference (5) for details). The density of the antigen in reversibly and irreversibly adsorbed form were 0.91 and 0.69 pmol/cm^2, respectively. The desorption rate constant of the antigen from the bead was 3.9×10^{-4} s^{-1}. The dissociation constant and the dissociation rate constant of the antibody (soluble scFv) from the antigen were 3.4×10^{-8} M^{-1} and 1.0×10^{-3} s^{-1}, respectively. The ratio of the effective antigen to the total antigen was 0.013.

An experiment was carried out under the same conditions as those used for the simulation to verify our kinetic model. As shown in Figure 2, the simulation results (line) agreed well with the experimental results (symbols). This shows that our kinetic model faithfully describes the biopanning process.

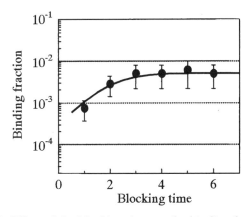

Figure 2. Effect of the blocking time on the binding fraction.
The line represents the simulation result and the symbols represent
the experimental data. The error bars indicate the standard
deviations. The parameters other than those described in the text are
$S/V=10^6$ cm$^2 \cdot l^{-1}$, *binding time=1 h, and washing time=1 h.*

Effect of Blocking Time on the Final Binding Fraction

The length of the blocking time has been set to be one or two hours in conventional protocols for biopanning. As shown in Figure 2, however, the

276

binding fraction increases with extension of the blocking process. After the coating process, the antigen adsorbed reversibly on the solid phase continues to desorb because the library solution and the washing buffer don't contain the antigen. When the blocking time is short, the degree of the desorption during the binding process is large because the desorption is a first-order reaction (Figure 3). Whereas the ratio of the effective antigen to the total antigen on the solid phase is only 0.013, most of the desorbed antigen molecules in solution are effective for phage binding because they rotate freely. Since the phage antibody recognizes the antigen both on the solid phase and in solution, the fraction of the phage antibody bound to the antigen on the solid phase decreases (the phage antibody bound to the antigen molecules in solution will be removed by washing). The extension of the blocking process reduces the degree of the desorption during the binding process because the fraction of the antigen in reversibly adsorbed form on the solid phase decreases, resulting the increase in the binding fraction.

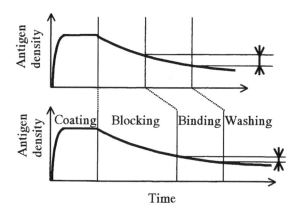

Figure 3. Effect of extension of the blocking time on the degree of desorption of the antigen during the binding process.

A Recommended Solid Support

As mentioned above, desorption of antigen molecules leads to a decrease in the biopanning output. Therefore, antigens should be immobilized tightly on the solid phase to avoid the loss of positive clones although the degree of the desorption can be reduced by extension of the blocking process. Considering experimental conveniences, a procedure combining the use of streptavidin beads and biotinylation of the antigen is one of the best choices for the following reasons.

(i) Since the dissociation rate constant of biotin from streptavidin is extremely low (2.4×10^{-6} s^{-1})([12]), desorption of antigens from the solid phase is completely negligible when streptavidin is covalently immobilized on the bead and the antigens are biotinylated covalently.

(ii) It is known that a significant fraction of antigens are denatured when the antigens are immobilized on solid phases by passive adsorption ([13–15]). The isolated antibodies might recognize the denatured antigens but not the native ones. When the antigens are biotinylated with a linker of suitable length and the conjugates are reacted with the streptavidin, the denaturation of antigens by the immobilization can be minimized.

(iii) Streptavidin beads with a high binding capacity (1 nmol biotin conjugates/mg beads) are commercially available. In addition, the β value of antigens immobilized on the streptavidin beads is expected to be much higher than that immobilized by passive adsorption ([15]). Therefore, a high effective antigen concentration can be achieved using streptavidin beads as the solid support.

(iv) Waste of precious antigens can be minimized because most of the biotinylated antigen molecules react with the streptavidin beads, while passive adsorption requires a high concentration of antigens.

Effect of the Effective Antigen Concentration on the Final Binding Fraction

When an antigen is biotinylated and is immobilized on streptavidin beads, the density of the antigen on the solid phase in reversibly adsorbed form is considered to be zero. Therefore, Eq. (2) can be simplified to be

$$F = \frac{C^{\mathit{eff}}}{K_d + C^{\mathit{eff}}} \exp(-k_{\mathit{off}} t_w) \tag{3}$$

where C^{eff} is the effective antigen concentration ($= \beta \Gamma_0^{\mathit{in}} S/V$). The dissociation constant is the ratio of the dissociation rate constant to the association rate constant ($K_d = k_{\mathit{off}}/k_{\mathit{on}}$). It was reported that the association rate constant (k_{on}) of antibodies is almost constant under constant temperature and viscosity because the association rate constant is mainly dependent on the diffusion rate of antibody molecules ([16, 17]). Figure 4 shows the effect of the effective antigen concentration on the binding fraction of the antibodies with the K_d values from 10^{-6} to 10^{-9} M when the k_{on} value of each antibody is assumed to be 10^4 M^{-1}s^{-1}.

When the immunotube (Nalgen Nunc) is used as the solid support, the effective antigen concentration was calculated to be 1.5×10^{-10} M (5). In the case that polystyrene dishes or ELISA plates are used as the solid support, the effective antigen concentration is expected to be lower than that in the case of the immunotube. The effective antigen concentration in the case of the beads used in this study was estimated experimentally to be 2.0×10^{-8} M (5). In the case that the antigen is biotinylated and immobilized on a streptavidin bead (for example, streptavidin magnesphere paramagnetic bead, Promega), the total antigen concentration is calculated to be 1.4×10^{-6} M based on the manufacturer's data when 0.6 mg of the bead is suspended in 0.2 ml. Since the β value in this case is expected to be at least more than 0.1 (15) (excepting the case that the epitope or its flanking region of the antigen is biotinylated preferentially), the effective antigen concentration should be at least above 10^{-7} M. Actually, we achieved more than 10^{-6} M of the effective antigen concentration when the biotinylated RNase A was immobilized on the streptavidin bead (data not shown). When 10^{12} phages of a typical phage antibody library with a diversity of 10^{9} are applied to biopanning and assuming that one-tenth of the phage molecules carry an antibody molecule (18), the population of each phage clone will be 10^{2}. To recover a positive phage clone, the binding fraction must be above 10^{-2} in this case (the gray area in Figure 4).

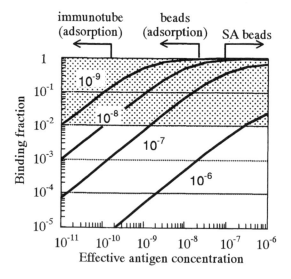

Figure 4. Effects of the effective antigen concentration on the binding fraction ($t_w = 5$ min).

When the immunotube is used as the solid support, only phage clones displaying a strong antibody with the K_d values below 10^{-8} M will be recovered. In contrast, when the beads with a large surface area are used as the solid support, even phage clones displaying a weak antibody with the K_d value around 10^{-6} M will be recovered. Since there is no guarantee that a library contains phage clenes displaying a strong antibody against an antigen of interest, researchers should immobilize their antigens on streptavidin beads at least in the first round of biopanning.

The Causes and Solutions for the Problems

Poor reproducibility

Table I shows the outputs of the phage antibody when the phage antibody is assumed to be contained in the typical library described above. When the immunotube is used as the solid support, the output is calculated to be only 0.054. This means that the phage antibody can be isolated from the library only once in nineteen trials. This should be the cause of the "poor reproducibility". In contrast, when the beads are used as the solid phase, the clone will be recovered because the outputs are more than one.

Table I. Comparison of the Outputs

Solid support	Immunotube [a]	polystyrene bead [b]	Streptavidin bead [c]
Immobilization	adsorption	adsorption	streptavidin-biotin
V (ml)	4.0	0.2	0.2
S (cm^2)	14 [d]	2×10^2 [d]	$6-9 \times 10^2$ [d]
C^{eff} (M)	1.5×10^{-10}	2.0×10^{-8}	1×10^{-7} [e]
Output	0.054	6.0	23

Eq. (3) was used for the calculation of the output in the case of the streptavidin bead. In the other cases, Eq. (2) was used. Parameters other than those described in the text or the table are blocking time=6 h, binding time=2 h, washing time=20 min, and the population of the phage antibody=100.

[a] Nalgen Nunc, 444202.
[b] Polystyrene paramagnetic microparticles (2.6 mg), Polysciences, Inc.
[c] Streptavidin magnesphere paramagnetic bead (0.6 mg), Promega.
[d] Calculated based on each manufacturer's data.
[e] A higher concentration than this value will be expected actually (see text).

High background

In the washing process, an washing operation (adding a washing buffer and discarding it) is repeated several times to remove nonspecific phages. By this operation, nonspecific phages are removed by a dilution effect (7). It is known experientially that the output of the nonspecific phages decreases rapidly by the dilution effect in the initial period of the washing process but slowly in the later period and that a significant number of nonspecific phages still binds to the solid phase even after extensive washing (Figure 5, broken line). However, since the dissociation of antibody from the antigen is a first-order process, the dissociation of the specific phages is dependent on the total incubation time in the washing buffer and not on the number of the washing operation. That is to say, when logarithm of the output is plotted against the washing time, the output of a specific phage decreases linearly (Figure 5, solid lines). Desorption of the specific phages from the antigen on the solid phase must be distinguished from the dilution of the nonspecific phages. Therefore, we defined the washing time as the length of the incubation time when the solid phase is in the washing buffer including the time required for collection of the beads by a magnet during the washing step.

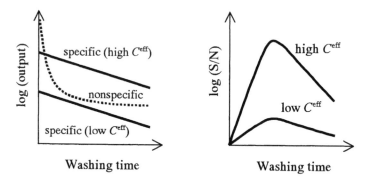

Figure 5. Time course of the output of the specific and nonspecific phages and the S/N (specific/nonspecific).

When the effective antigen concentration is low, the output of the specific phages is low, resulting that the ratio of the output of the specific phages to that of the nonspecific phages (S/N) is low. This should be the cause of the "high background". When the biopanning is performed under a high effective antigen concentration using the beads with a large surface area, the initial output (the intercept the left figure) of the specific phages increases. Although the output of the nonspecific phages may also increase under the high effective antigen concentration, the degree of the increase in the output of the nonspecific phages should be much lower than that of the specific phages. That is to say, when the effective antigen concentration is high, the S/N is high. This observation is supported experimentally by McConnel's study (*19*). They reported that much more positive clones can be obtained using the bead as the solid support comparing with the case using the immunotube. In a conclusion, researchers should use a solid support with a large surface area like bead to improve the S/N of the output.

Low affinity

It is known that the extension of the washing time is effective for isolation of phage antibodies with high affinity against their antigen (*9, 20*). However, the importance of the washing time has not mentiond in most of conventional protocols and most of researchers have not used this knowledge in their biopanning. This might be bacause the appropriate washing time could not be designed based on the previous kinetic model (Eq. (2)) in which the total antigen concentration was used for calculation of the binding fraction. In contrast, it can be discussed quantitatively based on our model because the effective antigen concentration has been determined. The effect of the washing time on the binding fraction of a phage displaying a weak antibody ($K_d=10^{-7}$ M, $k_{off}=10^{-3}$ s^{-1}) or a strong antibody ($K_d=10^{-9}$ M, $k_{off}=10^{-5}$ s^{-1}) was simulated (Figure

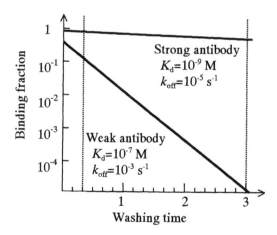

Figure 6. Effects of the washing time on the binding fraction ($C^{eff}=10^{-7}M$)

6). After washing for 20 minutes, the binding fractions of the strong and the weak phage are calculated to be 0.98 and 0.15, respectively. Since the fraction of the weak phages in antibody libraries is much higher than that of the strong phages (probability of the weak antibodies is expected to be around 10^3-fold higher than that of the strong antibodies based the results of Lancet et $al.$ (21)), most of the output phages may display weak antibodies. This should be the cause of the "weak affinity". In contrast, after washing for three hours, the binding fractions of the strong and the weak phage will become 0.89 and 10^{-5}, respectively, resulting that most of the output are strong phages. However, it must be noted that the extension of the washing time decreases the binding fraction of positive clones. In the first round of biopanning, researchers shold not extend the washing time even when they employ a streptavidin bead as the solid support because the popuration of each phage clone is low as mentioned above. The following procedure is recommended for isolation of strong clones.

1. Estimate the popuration of each phage clone (N) in the second (or further) round of biopanning (the ratio of the input for the second round to the output of the first round).
2. Assume the K_d value of an antibody that researcher wishes to isolate.
3. Calculate the k_{off} value of the antibody assuming $k_{on}=10^4$ $M^{-1}s^{-1}$.
4. Assume that the effective antigen concentration is 10^{-7} M in the case of employing the streptavidin-biotin system. Alternatively, estimate it based on the biological activity of the antigen of interest immobilized on the beads.
5. Caluculate the appropriate washing time which satisfies the unequal equation $F \cdot N > 1$ based on the Eq. (3).

References

1. Marks, J.D.; Hoogenboom, H. R.; Bonnert, T. P.; Macafferty, J.; Griffiths, A. D.; Winter, G. By-passing immunization: Human antibodies from V-gene libraries display on phage. $J.$ $Mol.$ $Biol.$ **1991**, *222*, 581-597.
2. Scott, J. K.; Smith, G. P. Searching for peptide ligands with an epitope library. $Science$ **1990**, *249*, 386-390.
3. Rebar E. J.; Pabo, C. O. Zinc finger phage: Affinity selection of fingers with new DNA binding specificities. $Science$ **1994**, *263*, 671-673.
4. Parmley, S.F.; Smith, G. P. Antibody-selectable filamentous fd phage vectors: Affinity purification of target genes. $Gene$ **1988**, *73*, 305-318.
5. Zhuang, G; Katakura, Y.; Furuta, T.; Omasa, T.; Kishimoto, M.; Suga, K. A kinetic model for a biopanning process considering antigen desorption

and effective antigen concentration on a solid phase. *J. Biosci. Bioeng.* **2001**, *91*, 474-471.

6. Katakura, Y.; Kobayashi. E.; Kurokawa, Y.; Omasa, T.; Fujiyama, K.; Suga, K. Cloning of cDNA and characterization of ˙Anti-RNase A monoclonal antibody 3A21. *J. Ferment. Bioeng.* **1996**, *82*, 312-314.

7. Mandecki, W.; Chen, Y. C. J.; Nelson, G. A mathematical model for biopanning (affinity selection) using peptide libraries on filamentous phage. *J. Theor. Biol.* **1995**, *176*, 523-530.

8. Kretzschmar, T.; Zimmermann, C.; and Geiser, M. Selection procedures for nonmatured phage antibodies: a quantitative comparison of optimization strategies. *Anal. Biochem.* **1995**, *224*, 413-419.

9. Levitan, B. Stochastic modeling and optimization of phage display. *J. Mol. Biol.* **1998**, *277*, 893-916.

10. Proteins at interfaces; Comstock. M. J., Eds.; American Chemical Society, Washington, DC, 1987; p 1-33.

11. Malmsten, M. Ellipsometry of protein layers adsorbed at hydrophobic surfaces. *J. Colloid Interface Sci.* **1994**, *166*, 333-342.

12. Piran, U.; Riordan, W. J. Dissociation rate constant of the biotin-streptavidin complex. *J. Immunol. Methods* **1990**, *133*, 141-143.

13. Butler, J. E.; Navarro, P.; Sun, J. Adsorption-induced antigenic changes and their significance in ELISA and immunological disorders. *Immunol. Invest.* **1997**, *26*, 39-54.

14. Friguet, B.; Djavadi, O. L.; Goldberg, M. E. Some monoclonal antibodies raised with a native protein bind preferentially to the denatured antigen. *Mol. Immunol.* **1984**, 21, 673-677.

15. Butler, J. E.; Ni, L.; Nessler, R.; Joshi, K. S.; Suter, M.; Rosenberg, B.; Chang, J.; Brown, W. R.; Cantarero, L. A. The physical and functional behavior of capture antibodies adsorbed on polystyrene. *J. Immunol. Methods* **1992**, *150*, 77-90.

16. Goldbaum, F. A.; Cauerhff, A.; Velikovsky, A.; Llera, A.; Riottot, M. M.; Poljak, R. J. Lack of significant differences in association rates and affinities of antibodies from short-term and long-term responses to hen egg lysozyme. *J. Immunol.* **1999**, *162*, 6040-6045.

17. Foote, J.; Eisen, H. N.: Kinetics and affinity limits on antibodies produced during immune responses. *Proc. Natl. Acad. Sci. USA.* **1995**, *92*, 1254-1256.

18. Bradbury, A.; Persic, L.; Werge, T.; Cattaneo, A. Use of living columns to select specific phage antibodies. *Bio/technology* **1993**, *11*, 1565-1569.

19. McConnell, S. J.; Dinh, T.; Le, M. H.; Spinella, D. G. Biopanning phage display libraries using magnetic beads vs. polystyrene plates. *Biotechniques* **1999**, *26*, 208-214

20. Hawkins, R. E.; Russell, S. J.; Winter, G. Selection of phage antibodies by binding affinity. Mimicking affinity maturation. *J. Mol. Biol.* **1992**, *226*, 889-896.
21. Lancet, D; Sadovsky, E.; Seidemann, E. Probability model for molecular recognition in biological receptor repartries: Significance of the olfactory system. *Proc. Natl. Acad. Sci. USA* **1993**, *90*, 3715-3719.

Chapter 23

Open Sandwich Selection: Selection of Human Antibody Fragments Using the Mechanism of Fv Fragment Stabilization in the Presence of Antigen

Kouhei Tsumoto, Hideki Watanabe, and Izumi Kumagai

Department of Biological Engineering, Graduate School of Engineering, Tohoku University, Abba-yama 07, Aoba-har, Sendai 980–8579, Japan

Selection and preparation of antibody fragments have been intriguing for various fields, e.g. immunotherapy, using phage-display system. However, a lot of researchers have been experienced several technical troubles. Most antibody variable domain (Fv) fragments of immunoglobulins readily dissociate under physiological conditions, and are usually stabilized in the presence of their antigens. On the basis of this mechanism, we have proposed the "open sandwich (OS)" method for determining antigen concentration. The OS method can also be applied to the selection of novel antibody fragments. Here we will suumarized the selection of human antibody fragment from naive library using the OS-selection method. Several antibody VH fragments specific for erythropoietin receptor (EPO-R), hen egg-white lysozyme, BSA, and human Fc could be enriched.

A major goal of protein engineering is the design of proteins with novel and/or desired functions, e.g. specific recognition of target molecules.

© 2002 American Chemical Society

Antibodies can be regarded as a protein engineering system perfected by nature for the generation of a virtually unlimited repertoire of complementary molecular surfaces [1,2].

Phage display has been widely used as a tool for evolutionary engineering, e.g. selection of peptides and proteins specific for target molecules [3,4]. However, a lot of technical problems have been pointed out, e.g. efficiency of the surface display, stability of genes during selection, removal of non-specific binding clones.

Most Fv fragments of immunoglobulins readily dissociate under physiological conditions. However, Fv fragments are usually stable in the presence of their antigens. On the basis of this mechanism, we have proposed the "open sandwich (OS)" method for determining antigen concentration [5]. Sensitivity of the method is relatively high, and the background is significantly lower than the usual method. Thus, the OS method can also be applied to the selection of novel antibody fragments [6] (Figure 1). We have recently reported enhancement of the affinity of an anti-hen lysozyme antibody, HyHEL-10, toward a mutated antigen, other lysozymes, which was performed by saturation

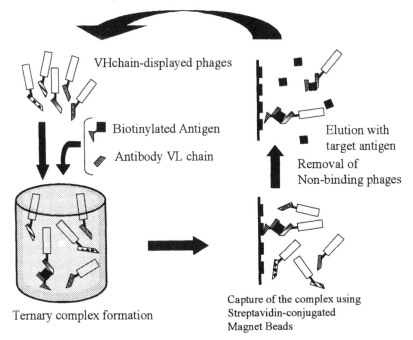

VHchain-displayed phages

■ Biotinylated Antigen

▨ Antibody VL chain

Elution with
target antigen
Removal of
Non-binding phages

Ternary complex formation

Capture of the complex using
Streptavidin-conjugated
Magnet Beads

Figure 1 Open sandwich selection

mutagenesis into four residues in complementarity determining region 2 of heavy chain (CDR-H2) followed by selection with affinity for TEL [7]. Using the OS-selection, several clones with converted specificity could be obtained. Thus, the selection method would be applicable to the novel antibody fragment from naive human antibody library.

Here we would describe the selection of antibody fragment from naive library using the OS-selection method. To date, several human antibody VH fragments specific for human soluble erythropoietin receptor (s-hEPO-R), bovine serum albumin (BSA), and human Fc could have already been selected, and here the case of s-hEPO-R has been described.

Materials and Methods

Materials

Construction of human antibody library of variable regions of the heavy and light chains (VH and VL) were described elsewhere [8]. Hen egg-white lysozyme was from Seikagaku-kogyo (Tokyo, Japan). Soluble erythropoietin receptor (s-hEPO-R) were prepared as described elsewhere. The synthesis of oligodeoxyribonucleotides for polymerase chain reactions (PCRs) was consigned to Nippon Gene Research Laboratories (Sendai, Japan). All enzymes for genetic engineering were obtained from TaKaRa Shuzo (Kyoto, Japan), Toyobo (Osaka, Japan), Roche (Mannheim, Germany), and New England Biolabs (Beverly, MA).

Preparation of Phage and Selection of Phage Antibody Libraries

Phage was recovered by PEG-precipitation [5,9] and resuspended in 500μl phosphate buffered saline (PBS) per 20ml of culture. In order to select the phage antibody libraries, streptavidin magnet beads were used. In the first step in the procedure, one hundred microliters of phage-VH were mixed with soluble biotinylated target antigens and certain purified human VL fragment at room temperature for 1 hour. Next, 50 μl of streptavidin-conjugated paramagnetic beads (Promega) was added and rapidly mixed. After an incubation period of 10 minutes, a magnet (Promega) was used to select out the beads and attached phage. The beads were washed stringently 5 times with 1 ml of PBST, and 5 more times with 1 ml of PBS. The bound phages were firstly eluted with 500 μl of PBS containing 1 μg target antigen, followed by second elution with 100 mM Glycine (pH 2.0) containing 200 mM NaCl, rapidly neutralizing with 100 mM Tris-HCl (pH 8.5) containing 500 mM NaCl. Each elute was reinfected into early log-phase *E. coli* JM109. After 1 hour, 1000 μl of the culture was plated and incubated at 37°C for overnight. Clones on the plates were transferred into the LB culture containing 100mg/l Ampicillin, and infected with helper phage, M13KO7, and incubated overnight at 37°C. The phage was prepared by centrifugation and PEG-precipitation and was subjected to the next selection stage. Following the four stages, the clones on plates were cultured and phage-VH was prepared according to the method described previously [5].

Characterization of Selected Phage-VH Fragments

Di-deoxy reactions for DNA sequencing were performed using an auto-read sequencing Kit (Pharmacia) according to the manufacturer's recommendations. The analysis of DNA sequences was performed using ALF express auto-read sequencer (Pharmacia). Open sandwich ELISA was performed according to the method described by Ueda et al. [5]. One hundred microliters of 10 μg/ml human VL fragment in PBS was applied to each well of the microtiter plates and incubated for 1 hour at room temperature. After removing the solution, 200 μl SuperBlock (Piearce) was added to each well and incubated for 1 hour at room temperature. After discarding the buffer, 100 μl of VH-phage, which had been mixed with target antigens and prediluted with 1 volume of binding buffer 30 minutes before, was added to each well and incubated for 1 hour at room temperature. After washing twice with PBS-T, 100 μl of 5000 times-diluted horseradish peroxidase (HRP)-anti M13 (Pharmacia) was added and the mixture was incubated for 1 hour. After three more washings with PBS-T, 200 μl of 50 mM sodium succinate buffer containing 10 mg/ml ABTS and 0.01% H_2O_2 was added and incubated for 1 hour. Absorbance was measured at 415 nm using microplate reader (Bio-rad, type 550), using 630 nm as the control.

Expression and purification of Fv fragment selected

The expression vectors for VH and VL (designated as pUT-VH, and pUT-VL, respectively) were constructed as described previously. In brief, *E. coli* BL21(DE3) cells [10] transformed with pUT-VH or pUT-VL were cultured at 28 °C in LB containing 100 μg/ml ampicillin in shaking flasks to an optical density of 0.8 at 600 nm and induced by addition of 0.1 mM IPTG. Culture growth was continued for 12–16 h, raising the temperature to 37 °C. Cells were harvested and resuspended in 10 mM Tris–HCl (pH 8.0) and 1 mM EDTA. The suspension was sonicated for 15 min at 150–170 W and centrifuged for 20 min at 9000 rpm. The pellet obtained was dissolved in 20 mM Tris–HCl (pH 7.9), 0.5 M NaCl, and 5 mM imidazole containing 6 M guanidinium hydrochloride (GuHCl) and applied to His•Bind Resin (Novagen) preequilibrated in the same buffer as used in pellet solubilization. The (His)$_6$-tagged variable region fragments were eluted with the buffer containing 60 mM imidazole. The molecular weights of purified VH and VL were confirmed on a 15% SDS–PAGE in the same buffer system described by Laemmli [11].

Refolding and purification of Fv fragments selected

Purified variable region fragments in denaturing conditions were diluted to a concentration of about 7.5 μM and reduced by addition of β-mercaptoethanol (β-ME). In the case of Fv fragments whose domains were expressed separately, VH and VL were mixed stoichiometrically at the same concentration. Procedures for the refolding of insoluble antibody fragments described by Tsumoto et al. [12] were followed. After a short time, the concentration of GuHCl in the dialysis buffer was phased down (3, 2, 1, 0.5, and 0 M), and an oxidizing reagent (glutathione, oxidized form, Sigma, St. Louis, MO) and 0.4 M L-Arg were added during the final dialysis stage. The supernatants containing refolded proteins were concentrated to approximately 10 μM using a Centriplus YM-10 ultrafiltration membrane followed by a Centricon YM-10 membrane (Millipore, Bedford, MA). The Fv fragment obtained was purified on a Superdex 75pg (16 x 400 mm) equilibrated with 50 mM Tris-HCl (pH 7.5) containing 200 mM NaCl, and finally dialyzed overnight against 50 mM phosphate buffer (pH 7.2) containing 200 mM NaCl.

Determination of binding affinities toward target antigen

Kinetic and equilibrium constants for the interactions between target antigens and the variable region fragments selected were measured with an IAsys Auto+ optical biosensor (Affinity Sensors, UK). The s-hEPO-R was immobilized on the surface of a carboxymethyl dextran (CMD) cuvette using EDC/NHS chemistry. Variable region fragments at various concentrations in PBS were added onto the target antigen-immobilized cuvette, and the interactions were monitored instantaneously by IAsys 3.0 software. The resulting association curves were processed as described previously [13]. In brief, the association rate constant (k_{ass}) was obtained by plotting the slope, k_s, of the dR/dt vs R plot for all concentrations of the selected Fv against the Fv concentration. The slope of this new plot is the k_{ass}, and the intercept, in theory, is the dissociation rate constant (k_{diss}).

Results and Discussion

Selection of clones with high affinity for s-hEPO-R by OS-selection

The human antibody VH libraries were subjected to a selection on the basis of the mechanism of Fv fragment stabilization under coexistent biotinylated antigens (i.e. OS-selection) using Streptavidin magnet beads (Figure 1). Phage-VH was mixed with soluble target antigens and purified certain human VL chain, forming ternary complex by incubation for an hour at room temperature. The resulting ternary complex was captured by streptavidin-conjugated paramagnetic beads, followed by selection using a magnet. The beads were then washed meticulously, and the adsorbed phage particles were recovered by competitive elution, followed by acid-elution. Then, the eluted phages were reinfected into early log-phase *E. coli* cells. The prepared phage was subjected to the next selection stage. ELISA analysis [5] using selected phages and soluble VL fragment clearly indicated that clones with increased affinity toward target antigen were enriched, and the number of clones increased during selection (data not shown). After three rounds of panning, clones on the culture plates were randomly picked up, and the DNA sequences of the clones were determined (Figure 2). ELISA analyses have indicated that the clones selected showed a significant ELISA signal for each antigen (data not shown).

Clone	CDRH1	CDRH2	CDRH3
3B1	NYPIS	GIIPVLGIQNDAQKFQD	GSGGDTGYF------DL
3C2	GYEMN	YISSSGSTIFYADSVKG	DHMTTVTPF------DS
3G5	DYYRT	NIYN-SGINKYNPSLES	WGPGYFNGGSCYSF-DH

Figure 2 CDR sequences of clones selected

Bacterial expression of selected Fv

Since most of the clones selected cannot be obtained with high yield by using secretory expression system, we attempted to construct the cytoplasmic expression system using pUT-E, which was constructed for expression of more than 100 mg/L of HyHEL10 scFv as insoluble proteins in cytoplasmic space [12], and the gene products have been obtained as inclusion bodies. VH, and VL, whose expressions were controlled by an inducible T7 promoter and terminator, fused with a $(His)_6$ tag at the carboxyl termini to facilitate purification. The *E coli* BL21(DE3) cells transformed with pUT-VH and pUT-VL were propagated to an OD_{600} of 0.8 at 28 °C and induced by addition of 0.1 mM IPTG. Insoluble materials including the recombinant proteins were obtained from the sonicated cell mixture as pellets and were solubilized by the denaturation buffer containing 6 M GuHCl.

Purification and refolding of variable region fragments selected

To purify the proteins, supernatants containing solubilized variable domain fragments were subjected to immobilized metal chelate affinity chromatography (IMAC). From the SDS–PAGE analysis, all variable region fragments were eluted at 200 mM imidazole in a very pure state.

The refolding methods of the variable region fragments described by Tsumoto *et al.* [12] were followed, based on the sequential decrease in the concentration of GuHCl from 3 to 0 M. In the case of Fv, the refolding was started in a 1:1 stoichiometric mixture of VH to VL under reducing conditions. Little aggregation was found in either refolding solution, indicating that the antibody fragments subjected to the refolding reaction were mostly present in the supernatant. We obtained 10 mg of Fv from 1 L of culture medium. The refolding reactions were also performed with VL and VH separately. Surprisingly, VL was almost completely refolded (more than 80% recovery), while some VH could not be refolded, but formed precipitates (data not shown). These results clearly indicate that the presence of the VL domain was required for the refolding of the VH domain, and interactions between the VH and the VL prompted the arrangement in the correct conformation, as suggested by Jäger and Plückthun [14].

Kinetics of the interactions between s-hEPO-R and the variable region fragments selected

The binding affinities of Fv fragments selected toward s-hEPO-R were estimated using an IAsys Auto+ optical biosensor. Bacterially expressed s-hEPO-R was immobilized onto the sensor surface since sufficient Lys in the MBP was expected to be used chiefly for linkage to CMD on the surface. The observed response (R) of the immobilized protein was 500 arc second, equivalent to a concentration of approximately 0.50 µM within the CMD

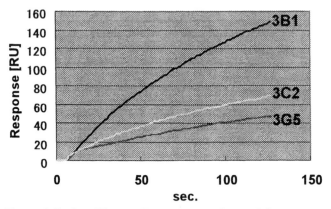

Figure 3 Surface Plasmon Resonance analyses of the interactions between clones selected and s-hEPO-R

matrix. The sensor surface was subjected to various concentrations of Fv fragments selected, and kinetic constants were calculated from the binding profiles as described under Materials and Methods. Figure 3 demonstrates the SPR profile of three clones selected (500 nM). Figure 4A shows the typical response for the antibody fragment binding (100–500 nM for 3B1 Fv) to the immobilized s-hEPO-R. Kinetic parameters obtained from the binding curves are shown in Table 1. The value of the affinity constant for 3B1, 3C2, and 3G5 estimated from k_{dis} / k_{ass} was 213, 369, and 612 nM, respectively.

Table 1 Kinetic parameters of the interactions between Fv fragments selected and s-hEPO-R

Clone	$k$on / $M^{-1}s^{-1}$	$k$off / s^{-1}	Kd / M
3B1	1.83×10^4	3.90×10^{-3}	2.13×10^{-7}
3C2	0.46×10^4	1.70×10^{-3}	3.69×10^{-7}
3B1	1.24×10^4	7.60×10^{-3}	6.12×10^{-7}

293

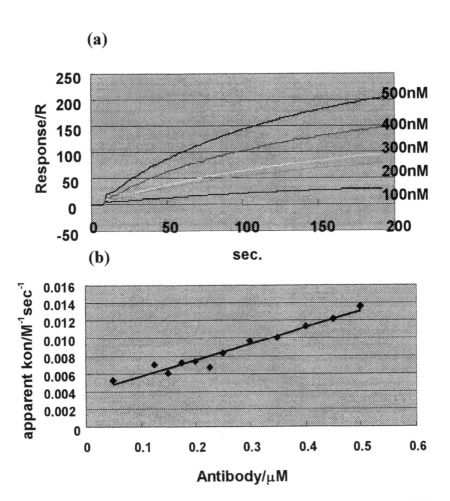

Figure 4 Surface Plasmon Resonance analyses of the interactions between 3B1 Fv and s-hEPO-R
(a) Concentration-dependency of the response. (b) kon-concentration plot based on dR/dt plot.

294

Applicability of OS-selection to preparation of human antibody fragments: Further perspectives

In vitro selection of antibody fragments (i.e. Fab and single-chain antibody, scFv) using phage display system has been reported by several groups. These systems, however, have some disadvantages as follows. 1) *Construction of library.* The library with 10^9 identical clones should be constructed. The technical problems for efficiency of transformation limits the library scale. 2) *Low display efficiency for proteins.* Phagemid display and reducing the molecule size displayed on phages would overcome the problem. 3) *Removal of non-specific binding clones.* 4) *Deletion of genes during selection.* The reason has not yet been clear, but the deletion of genes encoding antibody fragments is often observed during panning procedures. Utilization of Fv fragment would overcome these problems, and thus, stable and efficient selection can be performed by the OS-selection method reported here. Indeed, human antibody fragments specific for several antigens could be selected (Tsumoto, K. manuscript in preparation).

It has been known that the antibody molecules from naïve library have an affinity of *ca* 10^6 for target antigens [reviewed in Ref.3], and that enhancement of affinity can be achieved through affinity maturation *in vivo*. Recent technical progresses in random mutation using error-prone PCR [15] or cassette mutagenesis would make it feasible to mimic hypermutation *in vivo*, and affinity of the clones selected through OS-selection can also be enhanced.

It has been reported that some VH fragment can associate with certain VL fragments selected *in vivo*, and creating diversity into the six CDRs in certain antibody framework have been reported [16]. OS-selection would be applicable to these antibody libraries.

Acknowledgement

We would like to thank to Prof. Teruyuki Nagamune and Prof. Hiroshi Ueda of the University of Tokyo for their helpful suggestions with OS method, and also to Dr. Katsumi Maenaka of National Institute of Genetics and to Dr. Yoshiyuki Nishimiya of AIST for valuable suggestions with this work. This work is in part supported by the Industrial Technology Research Grant Program from the New Energy and Industrial Technology Development Organization (NEDO) of Japan.

References

1. Mariuzza, R.A. & Poljak, R.J. *Curr. Opin. Immunol.* **1993**, *5*, 50-55.
2. Padlan, E.A. *Adv. Protein Chem.* **1996**, *49*, 57-133.
3. Winter, G., Griffiths, A.D., Hawkins, R.E. & Hoogenboom, H.R. *Annu. Rev. Immunol.* **1994**, *12*, 433-455.
4. Forrer, P., Jung, S., and Plückthun, A. *Curr. Opin. Struct. Biol.* **1999**, *9*, 514-520
5. Ueda, H., Tsumoto, K., Kubota, K., Suzuki, E., Nagamune, T., Winter, G., Nishimura, H., Kumagai, I., & Mahoney, W.C. *Nature biotech.* **1996**, *14*, 1714-1718.
6. Tsumoto, K., Nishimiya, Y., Kasai, N., Ueda, H., Nagamune, T., Ogasahara, K., Yutani, K., Tokuhisa, K., Matsushima, M. and Kumagai, I. *Protein Eng.* **1997**, *10*, 1311-1318.
7. Nishimiya, Y., Tsumoto, K., Shiroishi, M., Yutani, K. and Kumagai, I. *J. Biol. Chem.* **2000**, *275*, 12813–12820
8. Nakamura, M., Watanabe, H., Nishimiya, Y., Tsumoto, K., Ishimura K., Kumagai, I., *J. Biochem.* **2001**, *129*, 209-212; Nakamura, M., Tsumoto, K., Ishimura, K., and Kumagai, I., *J. Immunol. Methods* **2001**, in press.
9. Maenaka, K., Furuta, M., Tsumoto, K., Watanabe, K., Ueda, Y., & Kumagai, I. *Biochem. Biophys. Res. Commun.* **1996**, *218*, 682-687.
10. Studier, F.W. & Moffatt, B.A. *J. Mol. Biol.* **1986**, *189*, 113-130.
11. Laemmli, U.K. *Nature* **1970**, *227*, 680-684.
12. Tsumoto, K., Shinoki, K., Kondo, H., Uchikawa, M., Juji, T. and Kumagai, I. *J. Immunol. Methods* **1998**, *219*, 119-129.
13. Ueno, T., Misawa, S., Ohba, Y., Matsumoto, M., Mizunuma, M., Kasai, N., Tsumoto, K., Kumagai, I. and Hayashi, H. *J. Virol.* **2000**, *74*, 6300–6308
14. Jäger, M. & Plückthun, A. *J. Mol. Biol.* **1999**, *285*, 2005-2019.
15. Knappik, A., Ge, L., Honegger, A., Pack, P., Fischer, M., Wellnhofer, G., Hoess, A., Wölle, J., Plückthun, A., and Virnekäs, B. *J. Mol. Biol.* **2000**, *296*, 57-86
16. Fromant, M., Blanquet, S. & Plateau, P. *Anal. Biochem.* **1995**, *224*, 347–353.

Author Index

Subject Index

A

Acetibacter xylinium, deep culture fermentation in bacterial cellulose production, 70

Acetone, solvent effects on silanization on silica gel surface, 96, 99*t*

Actinobacillus succinogens, efficient succinic acid producer, 31

Anaerobic metabolic pathways, *E. Coli*, 32-33*f*

Anaerobiospirillium succiniproducens, efficient succinic acid producer, 31, 33

Aequorea victoria, green fluorescent protein, biological probe, 236

Affinity selection process of phage display libraries, kinetic model, 271-284

α-Agglutinin, heterologous proteins on yeast cell wall display, 235-236

Amino acid sequences comparison
CDRL-1 peptide and VIP, 207
i41SL1-2 and light chain VIP cleaving antibody, 205*f*

3-Aminopropyltriethyoxysilane, effect on silanization on silica gel surface, 96, 98*f*

Amplex™ Red reaction by peroxidase with hydrogen peroxide, 213*f*

Amplification
PEP carboxylation flux, 34
pyruvate carboxylation flux, 34-36
selective transgene-expressing cells, possibility, 145-146

Animal cell systems, 3-4

Antibody light chain, catalytic activity, 200-207

Antigen concentration determination by open sandwich method, 285-295

Antigen concentration effect on final binding fraction, 277-279

Antigen-dependent proliferation, Ba/F3 transfectant, 144-145

Antigen-mediated genetically modified cell amplification, 140-152

Antioxidants, effects on cadmium ion toxicity, 169-174*f*

Antitumor effects in vitro, hybrid liposomes, 178-180

Antitumor effects in vivo, hybrid liposomes, 184-185

Antitumor mechanism, hybrid liposomes, 181-184

Apoptosis inhibiting genes and caspase inhibitors in mammalian cell survival, 190-199

Apoptosis inhibition, baculovirus-induced insect cell death, silkworm hemolymph effect, 157-160

Apoptosis signaling pathway, caspases activation, 192*f*

Apoptotic signal transduction by cadmium ion, 163-176

Artificial evolution techniques, 3

Aspergillus niger optimized enzyme system in paper sludge hydrolysis, 135

Avidin-biotin-immobilized liposomes, characterization, 261-263

T

Temperature effects on silanization on silica gel surface, 96, 99*t*

Therapeutic effect improvement, gene transduction efficiency, 149

Thermodynamic feasibility analysis in metabolic control analysis, 19

Tolerance, acquisition confirmation, 229, 231

Tolerance mechanism, cell surface-related, 228-229

Toluene used for bacteria removal from bacterial cellulose, 76-77

Torulopsis glabrata, metabolism and fermentation characteristics, 16-17

Transgene-expressing cells, selective amplification possibility, 145-146

2,3,7-Trichlorodibenzo-*p*-dioxin peptide ligand binding, 210-219 structure, 213*f*

2,4,6-Trichlorophenol in wastewater, continuous degradation, using membrane surface-liquid culture, 117-118

Trichoderma reesei
optimized enzyme system in paper sludge hydrolysis, 135
raw cellulase production, 122-123

Triggering operations, glutamate production in *Corynebacterium glutamicum*, 40

3-(Trimethoxysilyl) propylethylendiamine, effect on silanization on silica gel surface, 96, 98*f*

U

Unknown pathways identification, 20-21

V

Variable region fragments, purification and refolding, 291

Vector construction, display on yeast cell surface, 223-224*f*

Vitamin B_{12} fermentation, new pathways identification, 20-21

Volumetric efficiency improvements in rotating cylinder bioreactors by rectangular element use, 73-76*f*

W

Wastewater, continuous degradation, 2,4,6-trichlorophenol using membrane surface-liquid culture, 117-118

Water soluble materials in paper sludge, effect on enzyme activities, 123-124*t*

Western blotting
antigen-mediated genetically modified cell amplification, 143-144, 147
immune complexes detected by enhanced chemiluminescence, 165, 172

White-rot fungi, *Corilous versicolor*, laccase production, 109-110

X

Xylitol
biological production by *Candida tropicalis* and recombinant *Saccharomyces cerevisiae*, 53-67
production by chemical hydrogenation, xylose, 54
significance as alternative sweetener, 54

Xylitol productivity, by cell-recycle
fermentation using *C. tropicalis*, 60,
62-63*f*
Xylose, metabolic pathway in *Candida tropicalis*, 55*f*
Xylose reductase gene in fed-batch
fermenetation of recombinant, *S. cerevisiae*, 62, 64-65*f*

protein library on surface, 220-233
Yeast cell surface
combinatorial protein library
display, 222-225
green fluorescent protein display,
236-240
Yeast, surface display systems,
establishment, 222

Y

Yeast cell, model eukaryotic cells,
construction of combinatorial